T0305315

Economics and Environmental Change

Technology and Demographic Change

Economics and Environmental Change

The Challenges We Face

Clement A. Tisdell

Professor Emeritus, School of Economics, The University of Queensland, Australia

 Edward Elgar
PUBLISHING

Cheltenham, UK • Northampton, MA, USA

Published by
Edward Elgar Publishing Limited
The Lypiatts
15 Lansdown Road
Cheltenham
Glos GL50 2JA
UK

Edward Elgar Publishing, Inc.
William Pratt House
9 Dewey Court
Northampton
Massachusetts 01060
USA

A catalogue record for this book
is available from the British Library

Library of Congress Control Number: 2017933395

This book is available electronically in the **Elgar**online
Economics subject collection
DOI 10.4337/9781782549635

MIX
Paper from
responsible sources
FSC
www.fsc.org   FSC® C013604

ISBN 978 1 78254 962 8 (cased)
ISBN 978 1 78254 963 5 (eBook)

Typeset by Servis Filmsetting Ltd, Stockport, Cheshire
Printed and bound by CPI Group (UK) Ltd, Croydon, CR0 4YY

Contents

List of figures		vii
List of tables		ix
Preface		x
1	Economics and environmental change: an overview	1
2	Growing economic activity and environmental change: historical and general perspectives	9
3	Sustainable (economic) development: what is it? Is it desirable? Can it be achieved and if so, how?	32
4	Values, economic valuation, and the assessment of environmental and economic change	51
5	Social embedding: its nature and role in determining our economic and environmental future	78
6	Consumers' sovereignty – significant failures: why consumers' demands for environmental, human and animal protection are often unmet	94
7	Biological conservation and human-induced environmental change: contemporary socio-economic challenges	120
8	Climate change: general aspects, and alterations in energy sources and use as responses	149
9	Agriculture and environmental change, especially climate change: economic challenges	173
10	Marine ecosystems and global climate change: economic consequences, resilience and adjustment	202
Index		229

Figures

2.1 The possible implications of Ehrlich's equation as explained in the text   19
2.2 Stylized pattern of sea levels during the past 140 000 years   25
2.3 An illustration of the proposition that an acceleration in the regional loss of natural resources (due to uncontrolled environmental causes) increases the economic incentive to accelerate their use and reduces the economic worth of conserving them   28
3.1 An illustration that some development paths satisfying particular criteria for sustainable economic development are not the most socially desirable ones, as explained in the text   34
3.2 Hypothetical trade-off possibilities between the level of economic welfare of current generations and that of future generations used to illustrate problems discussed in the text   35
3.3 An impression of the range of different views about the need to conserve natural environments and resources in order to achieve sustainable economic development   37
3.4 A Venn diagram illustrating the three-pillar approach to choosing sustainable development strategies   40
4.1 An illustration of a valuation deficiency (explained in the text) when travel cost methods are used to estimate the aggregate economic surplus obtained from a site-specific attraction   59
4.2 An illustration of the possible conflict between the economic value of visiting a site and its off-site (economic) conservation value   60
4.3 An illustration of Bergson's analysis of optimal collective choices   69
5.1 A schematic representation of the 'typical' pattern of development in ancient times following the commencement of agriculture   86
6.1 An illustration of a case in which a food production method elevates health risks to consumers but nevertheless increases their economic welfare and the total economic surplus   105

6.2 An illustration that even if the demand for a product is perfectly elastic, the economic burden of meeting an increased food safety standard may not fall entirely on producers 107

7.1 Tree-like representation of factors that can influence the attitudes of individuals towards the conservation and management of biota 124

7.2 An illustration of two types of efficiency failings which an NGO may make in trying to conserve multiple species 133

9.1 Several (partially interdependent) socio-economic systems influence resource use: shortcomings in any of these can result in failure to achieve desired social goals 174

9.2 An illustration that it is usually not socially optimal to eliminate all external environmental costs associated with agricultural activity 179

9.3 An illustration of net agricultural revenue per ha in relation to climate conditions proxied by temperature 191

9.4 An example of a biological tolerance function that declines sharply once environmental conditions reach a particular threshold 194

10.1 An illustration of the importance of taking into account opportunity costs when evaluating the economics of changes in marine ecosystems 214

10.2 When all the effects of climate change (not only acidification) are taken into account, the impact on the aggregate receipts obtained by Norwegian fisheries may follow a different pattern from that predicted by Armstrong et al. (2012) 219

Tables

4.1 Possible (general) distortions in economic values obtained by
     applying stated preference methods   64
8.1 Median life-cycle emissions (g CO_2/kWh) of selected
     currently available commercial technologies for producing
     electricity   160
8.2 Average construction time and average lifetime of plant for
     supplying electricity using alternative technologies   161
9.1 Examples of adverse environmental externalities from
     agricultural activities   178
9.2 The set of agricultural technologies listed by Ringler et al.
     (2014) which, if adopted, are projected by their chosen
     modelling to increase maize, rice and wheat yields
     significantly by 2050   185
9.3 A list of expected environmental changes due to climate
     change and their consequences for agricultural production   189
10.1 An indication of the heterogeneity of marine ecosystems   204

Preface

Changes occur constantly in economic, social and biophysical systems. Sometimes these changes are slow and at other times rapid and unpredictable. Often they display significant interdependence, and there is strong evidence that this interdependence has increased substantially with economic growth and development. Furthermore, evolutionary changes in the structure and organization of societies (attributable to the nature of economic growth and development) have arguably made it more difficult for humankind to respond effectively to the growing threats to its well-being and its wishes for a better world. The primary purpose of this book is to explore and analyse the nature of the environmental challenges we now face, bearing in mind that a holistic approach is required. Therefore, while the main focus of this book is on economics, I have also found it essential to take account of social and biophysical factors; grapple with relevant philosophical issues; and pay attention to broad patterns of historical development.

Given the word limit for this book, I have had to be selective in my coverage. I decided to concentrate on fundamental analysis rather than being side-tracked by considering fine technical details. Otherwise, my main perspectives might have been blurred. Also, I do not repeat analyses that are readily available in the existing literature. Although the individual chapters in this book can be read independently, a better appreciation of my point of view is obtained by reading the whole book.

Every chapter of this book focuses on the relationship between economics and environmental change. This is even true (although it might not seem so at first sight) of the chapter on consumers' sovereignty because as a result of market development (the extension of markets, increased economic specialization and the evolution of longer, wider and more complicated production and distribution chains), consumers' sovereignty in relation to environmental, human and animal protection seems to have diminished.

The importance is stressed in this book of social embedding as a major factor influencing human values and decision-making. It limits the ability of humankind to respond to and to assess prospective and actual environmental changes. However, the nature of social embedding is not static. It alters in nature with the passage of time, as demonstrated in this text.

The motivation to write this book arose from a series of lectures I gave to postgraduate students in bioscience at Minzu University (the Central University of Nationalities) in Beijing. My lectures took account of the economic and environmental challenges faced by China as a result of its rapid rate of economic growth in recent years. The contents of these lectures are not reported here because I thought it better to adopt a global perspective. Nevertheless, I wish to thank Professor Dayuan Xue for inviting me to give these lectures and the students for their stimulating discussions which eventually enticed me to write this book.

My joint research with Professor Serge Svizzero (of the Université de la Réunion) focusing on the economic development of ancient economies and its implications for their social evolution and economic growth, has convinced me of the importance of considering the history of socio-economic evolution in order to better understand our current situation. Therefore, account is taken of this historical background, albeit quite generally. John Gowdy's recent publications on ultrasociality also rekindled my earlier interest in social embedding, as for example, evidenced in some of my previous works, for example, *Biodiversity, Conservation and Sustainable Development* (Ch. 6), *Human Values and Biodiversity Conservation*, and *Economics of Environmental Conservation* (Ch. 1). As a result I have given greater attention to this subject than I might have otherwise.

I am grateful to those with whom I have been able to discuss aspects of this work (for example, David Adamson), and to Tooraj Jamasb of Durham University for reading an earlier draft of Chapter 8 and providing me with helpful suggestions. Also, several scholars provided me with suggestions which were useful in preparing Chapter 10 and they are acknowledged at the end of that chapter.

Without access to the resources of the School of Economics at The University of Queensland, this book is unlikely to have been written. I appreciate the support of the Head of School, Professor Rodney Strachan, in this regard. Evelyn Smart did an essential and excellent job in completing the word processing of the manuscript and providing some research support. Thanks Evelyn.

My wife, Mariel, continued to be patient about my writing activities at varied spots throughout the house, usually not in my office where my desk is strewn with papers. Thanks, Mariel. Sparky, our 14-year-old small dog, accompanied me on my daily walks until this manuscript was nearly finished, but sadly, he departed before it was completed.

I hope you find this book of interest.

Clem Tisdell
Brisbane

1. Economics and environmental change: an overview

1.1 INTRODUCTION

The increased ability of humankind to control, transform and use the Earth's material resources to satisfy human wants is regarded by most people to be an important indicator of human progress. This ability has played a major role in the economic ascent of humanity, and has altered social structures and the nature of the organization of societies. Humankind has used this capacity to significantly transform natural environments. The means available to *Homo sapiens* to alter and manage natural environments are more powerful and pervasive than ever before and the environmental footprint of our species continues to escalate. Economic success has enabled massive rises in the level of global population to occur as well as an extraordinary increase in the level of aggregate consumption. As a result, the volume of natural resource use by humans continues to rise and natural environments (ecosystems) continue to undergo substantial and rapid change. Furthermore, *Homo sapiens* has well and truly out competed all other species for the use of natural resources.

The current situation poses several challenges for humanity. While not all the environmental changes occurring as a result of human economic activity are socially undesirable, several are. Some, such as human-induced climate change may lead to the future impoverishment of humanity on a global scale. This possibility is of particular concern because of the nature of social embedding in modern economies (see Chapter 5 of this book). Because of this embedding, the root causes of human-induced global warming may not be effectively addressed by politicians and the public or done so in a timely fashion.

In the foreseeable future, the major economic problem faced by humanity may not be shortages of raw materials but changes to natural and human-developed environments caused by economic activity. These changes include ecosystem deterioration (both of natural ecosystems and human-developed ecosystems, such as agro-ecosystems) and global environmental change caused by rising concentrations of atmospheric greenhouse gases (GHG). While this book gives most attention to these issues, it also addresses a

1

wider range of environmental and social challenges. In doing this, I combine economic, social and biophysical considerations wherever this is called for.

1.2   THE MAIN OBJECTIVES OF THIS BOOK

The main objectives of this book and the reasons for pursuing these can best be understood by considering the purposes of its individual chapters. A brief account of these therefore follows. However, it should be emphasized that the following account does not cover all the issues addressed in the individual chapters.

Human impacts on natural environments have existed for a very long period of time and have magnified in intensity and areal extent with the passage of time. The Agricultural Revolution and the Industrial Revolution escalated the magnitude of human effects on natural environments. However, the Agricultural Revolution was actually a relatively slow process and the Industrial Revolution had its roots in prior developments. The objective of Chapter 2 is to provide a very general historical overview of the 'stages' of economic development and how they have altered the relationship between humankind and natural environments.[1] These historical developments are related to basic models of the magnitude of anthropocentric environmental impact, such as the model of Ehrlich (1989).

Significant concerns have been expressed in recent decades about whether economic growth (and even current levels of human well-being) can be maintained given the growing impacts of humanity on natural environments and the increasing rate at which it is depleting natural resources. Given this concern, strong support has been expressed in principle for the desirability of achieving sustainable (economic) development. The aim of Chapter 3 is to show that the concept of sustainable (economic) development is not straightforward. It has varied interpretations and it is unclear whether current generations are prepared to make sufficient sacrifices to obtain it, even if it can be achieved. Moreover, views differ significantly on what is needed to achieve sustainable development. Realizing the objective of sustainable development is likely also to be quite difficult in modern societies because of the presence of social embedding. This is explained in Chapters 5 and 9.

We rely heavily on our values (including the metrics of economic valuation) to assess economic and environmental changes, even though our perceptions of reality are also very important. Often, our values also influence our perceptions of reality. Chapter 4 explores the worth for assessing environmental and economic change of varied types of values and the suitability of a wide range of methods of economic valuation for doing this.

Social embedding has a variety of manifestations, causes and consequences. It constrains human behaviours. For example, it limits the ability of communities to respond to unwanted human-induced environmental changes, for example, to the causes of human-induced climate change. The purpose of Chapter 5 is to outline the nature of social embedding in contemporary societies and to show its importance in influencing our economic and environmental future.

Competitive market systems have been extolled as an efficient method of economic organization because they are parsimonious in the amount of information needed by market participants to ensure that these systems operate efficiently in allocating resources (Hayek, 1948). A further purported advantage of these systems is that they ensure consumers' sovereignty. These views are, however, too simplistic. In modern economies, consumers do not have a sound knowledge of the safety, social and environmental consequences of their purchases or of their impacts on animal welfare, even though many consumers are concerned about these aspects. As a result, market failure occurs. The objective of Chapter 6 is to consider these failures and provide examples of these.

Today, human economic activity is the main cause of biodiversity loss (either directly or indirectly) and the scale and rapidity of this loss is such that it threatens economic sustainability and the level of human well-being. Consequently, Chapter 7 considers the significance of biological conservation for sustaining human well-being and investigates the shortcomings of markets, public bodies and non-government organizations (NGOs) as facilitators of biological conservation. In addition, commonly used economic techniques for valuing biological change are critically examined. This is done from a broad perspective rather than by focusing on minor details.

The consequences of rising concentrations of GHGs in the atmosphere leading to global warming, climate change and increased ocean acidification are a major contemporary concern. The main source of these gas emissions (principally, CO_2) is the use of fossil fuels as an energy source. While the ability of humankind to harness fossil fuels as sources of energy is and has been a major contributor to economic growth, it is now widely believed (on the basis of the available scientific evidence) to be leading to unsustainable global economic growth and could eventually have net negative impacts on human well-being. The purpose of Chapter 8 is to consider social impediments to reducing GHG emissions, to outline general biophysical relationships involved in the process of global warming and to give particular attention to the economic scope to increasingly meet energy demands by relying on sources such as solar and wind. The latter energy sources have significantly lower levels of CO_2 emissions than fossil fuel (such as coal) and also the use of biomass. The whole life-cycle of different methods of energy supply (plus other considerations) is taken into account. In addition, it is argued that there is limited scope for sequestering CO_2 by increasing the biomass of vegetation.

The globe's current level of human population is obligated to agriculture for survival and will become even more dependent on cultivars and domesticated organisms for its survival and welfare as the level of global population continues to increase. Today, it would only be possible to support a small fraction of the world's population purely by hunting and gathering. Agricultural production is an important source of environmental change and is sensitive to it. Therefore, in Chapter 9, it is proposed to consider the major influences of agriculture on environmental changes and the consequences of climate change for agricultural production, as well as adjustment issues. Furthermore, the challenges which agriculture faces in this century in meeting demands for increased agricultural production without contributing significantly to socially unacceptable environmental change are to be assessed.

Marine ecosystem services (marketed, unmarketed and partially marketed commodities) are estimated to have a very high aggregate economic value (Costanza et al., 2014). The purpose of Chapter 10 is to discuss estimates of this value, consider its sources and assess how it might be affected by rising atmospheric CO_2 levels giving rise to global warming, climate change and ocean acidification. Drawing on the available literature, particular attention is given to the estimated effects of rising GHG emissions on the value of production by the Norwegian fisheries and on the changing economic value of coral reefs. These values are highly sensitive to biophysical predictions about changes in marine ecosystems, and these estimates have major consequences for economically optimal adjustment policies. However, there is a problem because scientific predictions about the dynamics of changes in marine ecosystems as a result of global warming, climate change and ocean acidification sometimes differ significantly. Consequently, determining the economics of adjustment policies is complicated by this uncertainty. This is explored and the importance is stressed of taking into account opportunity costs in assessing optimal economic adjustments to alterations in marine ecosystems.

1.3   FURTHER INSIGHTS INTO THE CONTENTS AND APPROACH ADOPTED IN THIS BOOK

The current challenges we face in responding to major environmental changes, especially human-induced environmental changes, are considered basically from four different angles. These are:

- from a broad historical perspective;
- the need to take account of the limitations of available economic measures and methods for valuing environmental change;
- the imperfections of many scientific predictions about the nature course, and consequences of biophysical attributes altered by environmental change (such as increasing levels of atmospheric CO_2); and
- the presence of social embedding (of different types) as an impediment to humankind responding effectively to actual or predicted environmental changes, especially human-induced environmental change, including human-generated climate change.

Knowledge of each of these components is essential for appreciating our current environmental problems and our prospects for solving these. We cannot assume that *Homo sapiens* is entirely rational[2] and we know that the rationality of our species is limited both in relation to individual decision-making and group decision-making (Tisdell, 1996). Therefore, the ability of humankind to respond to unwanted aspects of environmental change and to manage this change where it is human-induced is likely to be imperfect, particularly so in relation to managing the emissions of GHGs. Just how imperfect remains to be seen. This book identifies some of the most important challenges which we confront in dealing with these problems and why this is so.

Social Embedding Could be our Most Challenging Constraint on Effectively Addressing Human-induced Climate Change

As a rule, most economists do not give much attention to the fact that individuals are embedded in social systems. This embedding affects the values of individuals, their economic prospects and their ability to bring about potential changes to the socio-economic system, and it limits the responses of society to its environmental challenges. Two polar views seem to exist in the literature. At one extreme, it is argued that individuals are so locked into our current market-dominated socio-economic system that this system has assumed the nature of a superorganism. Consequently, no individual or small group of individuals can influence societal developments. Those who think they can influence the evolution of the system may just be responding to values generated by the system itself. 'Solutions' to problems may also be system-generated. Consequently, decision-makers may be little more than puppets when it comes to reacting to some of the challenges of environmental change, including climate change. Even methods and measures of economic valuation are liable to be influenced by social embedding influences.

At the other end of the spectrum is the view that individuals are entirely rational and societies do have the power to shape their future rationally. Put differently, given the extent of their knowledge about their economic and environmental possibilities, societies can be expected to make socially optimal decisions.

The truth probably lies somewhere in between these extremes. Just where is difficult to specify. However, the importance of social embedding ought not to be underestimated, even though, as is argued in Chapter 5 of this book, it is not as strong as in the case of some ultrasocial species of animals, such as some species of ants, bees and termites. Furthermore, the reasons for the social embedding of humans differ from those resulting in the ultrasociality of some animal species. In addition, changes in the nature of human embedding in human communities have occurred in a much shorter period of time than in the case of ultrasocial animal species. These changes accelerated in the Holocene era, the last 12 000 years or so.

Social embedding in human societies is not static. Its rate of change has been much faster than in the case of ultrasocial animal species. Furthermore, social embedding in human communities is significantly shaped by economic developments, as is apparent from the historical record. Today, as is demonstrated in Chapter 7, social embedding makes it difficult for humankind to address effectively the causes of global warming.

Social embedding does not imply stationarity in the structure of human societies but limits the ability of humans to manage change: it adds an extra constraint to the ability of humans to determine their future. As societies become larger in size, as has happened historically, the ability of individuals to influence the decisions of others is reduced. In the present era, this is occurring at a time when the welfare of all is becoming more interdependent due to increased economic globalization and to the growing global environmental consequences of economic activity.

1.4   CONCLUDING COMMENTS

We need to consider the environmental challenges we face from a holistic point of view. This means that our biophysical knowledge and our socio-economic knowledge must be combined in assessing environmental changes, and in determining the scope for humankind to respond to these challenges and to make desirable decisions. Even though the main focus of this book is on economic considerations, I have endeavoured, wherever possible and appropriate, to relate these to social and biophysical features. In addition, I have tried to concentrate on major issues rather than be side-tracked by unnecessary details, which in my view would have clouded the presentation.[3]

The major conclusion (but not the only conclusion) which can be drawn from the work is that social embedding is the prime impediment to humanity responding effectively to many of its current environmental problems, especially climate change. The nature of contemporary social embedding is mainly a product of the dominance of the market system as a means of economic organization. This dominance is a 'logical' result of the historical process of economic evolution. The depth of this social embedding is reinforced because economic methods of valuation have co-evolved with the ascendency of the market system and reflect prevalent societal values and those of dominant social groups. While this may have short-run benefits, it also poses long-term dangers because of the failures to step outside the existing system and evaluate it.

NOTES

1. Actually the nature of economic development has not altered discretely by stages nor have societies developed in the same pattern (Svizzero and Tisdell, 2016). Nevertheless, we can identify significant economic events that have significantly changed the course of economic and social development. From this point of view, the posited economic stages of economic development assist our understanding of the historical processes that have shaped our current situation.
2. An interesting perspective on factors influencing human behaviour and their evaluation is provided by Karen Armstrong (2014, pp. 4–9). A problem with the rationality approach to human behaviour is that it seems to be impossible to establish the desirability of ultimate ends by means of rational argument. Also individuals differ significantly in their willingness or ability to exercise various forms of rationality. Emotional factors often swamp rational considerations. *Homo sapiens* is not an entirely rational animal. Furthermore, the extent to which an individual is rational is likely to vary depending on changing circumstances.
3. Some additional information about the motivation for writing this book is available in the Preface.

REFERENCES

Armstrong, K. (2014), *Fields of Blood: Religion and the History of Violence*, London: Bodley Head.
Costanza, R., R. de Groot, P. Sutton, S. van der Ploeg, S.J. Anderson, I. Kubiszewski, S. Farber and R.K. Turner (2014), 'Changes in the global value of ecosystem services', *Global Environmental Change*, **26**, 152–8.
Ehrlich, P.R. (1989), 'Facing the habitability crisis', *BioScience*, **39**, 480–82.
Hayek, F. (1948), *Individualism and Economic Order*, Chicago, IL: Chicago University Press.
Svizzero, S. and C.A. Tisdell (2016), 'Economic evolution, diversity of societies

and stages of economic development: a critique of theories applied to hunters and gatherers and their successors', *Cogent Economics & Finance*, **4**(1), 1161322.
Tisdell, C.A. (1996), *Bounded Rationality and Economic Evolution*, Cheltenham, UK and Brookfield, VT, USA: Edward Elgar Publishing.

2. Growing economic activity and environmental change: historical and general perspectives

2.1 INTRODUCTION

The prime objective of this chapter is to discuss generally the relationship between economic activity and environmental change in its broad historical setting. It is important to realize that several of the environmental challenges we are now facing have been faced by humankind in the past. Furthermore, it is instructive to know how previous societies have responded to these and to know the way in which our environmental challenges have evolved with the passage of time.

Human economic activity has had environmental impacts since the earliest of times and human beings have had to adjust to changes in their environments throughout their existence. Environmental changes include both those induced by human activities and those occurring independently of these. Initially, the impact of *Homo sapiens* on natural environments was relatively small, but subsequently these magnified as mankind developed new techniques of obtaining a livelihood, as populations increased and economic growth occurred. The levels of global population and the amount of economic activity continue to increase and new methods for altering natural environments are continually being devised. This has raised serious concerns about whether economic growth can be sustained globally.

Although economic thought has responded to this changing situation, it seems to have done so with a time lag and until quite recently, intermittently. It was not until the 1970s that the subject of environmental economics began to be taught in universities. Along with ecological economics (a later development), it remained for some time a 'fringe' subject. Opinions among economists remained divided (and still are) about its importance. This chapter should help to bolster the case for its importance.

This chapter is developed as follows: first, evolving historical relationships between human economic activity and the state of the natural environment are considered, taking into account stages of economic development starting from dependence on hunting and gathering and then the Agricultural Revolution, the Industrial Revolution and further developments. Subsequently, Ehrlich's equation for determining the general environmental effects of economic growth is introduced and assessed. The types of policies which it suggests for reducing the potentially adverse consequences of economic growth are examined. It is pointed out that the environmental Kuznets curve portrays a more favourable relationship between economic growth and the state of the natural environment. This is because it indicates that although environmental conditions may initially deteriorate with economic growth, they will improve if sufficient economic growth can be achieved. In crude terms, this may be taken to imply that the way to generate a better environment is to achieve and maintain a high level of economic growth. Consequently, income levels rise and the state of the environment improves, resulting in a win–win situation. While this is an attractive proposition, it is subject to many limitations.

Attention is then given to several environmental characteristics which have received little attention in the economic literature. It is noted that man-made environments are valuable and in many cases they are more valuable than natural ones. Secondly, the nature of environmental goods is quite diverse – some are private goods and others are shared, for example, some are club goods or are pure public goods. This has important implications for economic analysis. It has consequences for inequality in the enjoyment of environmental commodities. It is hypothesized that this inequality rises with income inequality.

From scientific evidence of past environments, we know that major environmental changes have occurred which were unrelated to human activity. Examples of such changes are given and human responses to these are considered. In regions where such changes occur, economic sustainability can be impossible. Possible human reactions (such as migration) to such changed environmental conditions are discussed. The following question is also posed: If natural environmental resources in a region are going to be lost due to uncontrolled environmental change, what difference ought this make to the exploitation (conservation) of these resources? A general answer is given.

2.2   EVOLVING HISTORICAL RELATIONSHIPS BETWEEN HUMAN ECONOMIC ACTIVITY AND THE NATURAL ENVIRONMENT

It is believed that *Homo sapiens* first appeared in Africa about 200 000 years ago and subsequently migrated to other parts of Earth (Renfrew, 2007).

The appearance and spread of our species has greatly transformed the natural world. This transformation has been closely associated with global economic growth and rising levels of human population. These variables tend to be interrelated; the former enables population levels to rise, but whether or not population increases depends on a variety of conditions. For example, Malthus's theory, that population can be expected to rise rapidly whenever per capita income exceeds subsistence level (Malthus, 1798), was found not to hold in countries that experienced substantial rises in per capita income and significant urbanization following the Industrial Revolution. Nevertheless, the level of global population has multiplied greatly since that time and continues to increase, even though it is predicted to peak in the second half of this century. This increased global population combined with desires for higher per capita income (particularly in lower-income countries) is likely to result in further substantial transformation of natural environments. This process is unlikely to stop once global population peaks, since it is probable that those on lower incomes will seek higher incomes, as well as many who are much better off. Consequently, questions about the sustainability and desirability of continuing economic growth are likely to persist. The impact on the quality of life of economic growth and the composition of production and consumption will also remain a concern.

The development of new techniques or methods for obtaining provisions to support human populations has played an important role in overcoming economic scarcity, transforming natural environments and enabling humankind to control or manage nature. In general, new methods of economic production combined with increased capital investment have resulted in a substantial reduction in economic scarcity. Nevertheless, they have not been the only important contributors to this. For example, institutional changes have also been very important. These include the use of money and the evolution of market systems. Complex interdependent mechanisms have resulted in the economic and environmental situation in which we now find ourselves.

Stages of Economic Development

In the eighteenth and nineteenth centuries, it was not uncommon to divide the evolutionary pattern of the economic development of human societies into a number of discrete stages or periods. Adam Smith, for example, envisaged four stages in economic development (Brewer, 2008; Meek et al., 1978) each of which reflected the primary means of obtaining a livelihood. In Smith's schemata, human societies initially relied on hunting and gathering to survive, then pastoralism was practised followed by agriculture (the growing of crops), and in the last stage, commerce. He characterized this last stage as involving much trade (including foreign trade), manufacturing, considerable division of labour and increased economic specialization. This stage he also associated with growing urbanization.

While the earliest human societies did depend exclusively on hunting and gathering for their survival, patterns of subsequent economic development varied (Svizzero and Tisdell, 2015a; 2016). Different parts of the world transited to agropastoralism and to commerce as substantial economic activities at different times and with varying rates of change. A few communities remained in the hunting and gathering stage until comparatively recently. As economic development occurred, different modes of earning a living usually co-existed, and even today products obtained from hunting and gathering still contribute to the consumption of most people, even though their proportionate contribution continues to decline. It is quite low in higher-income countries. In these countries, wild-caught fish comprise the main component of consumption of goods obtained from the wild but as aquaculture develops, this proportion is also declining.

Even in ancient societies, some of the attributes which Smith associates with the age of commerce were present. Many hunting and gathering tribes did engage in long distance trade and several agriculturally based societies (such as that of ancient Egypt) did likewise. Long distance trade was also an important activity for the Minoans and Mycenaeans (Svizzero and Tisdell, 2015b), and especially for the Phoenicians (Markoe, 2000) as well as the Romans. There was some division of labour. However, these attributes became more marked as economic development occurred.

One may also argue that three types of major economic revolution have occurred during human existence and that these ultimately resulted in a substantial increase in global population and in production. These are:

- The shift from hunting and gathering to agriculture. Childe (1936 [1966]) considers this to be the First Economic Revolution. However, in reality, it involved a relatively slow evolutionary process. It began at different times in different parts of the world, initially beginning around 10 000 years ago, after the Earth emerged from the last major Ice Age (Svizzero and Tisdell, 2014b).
- The shift from agriculture to manufacturing as a major economic activity marked the beginning of the Industrial Revolution (the Second Economic Revolution). This commenced in Britain in the last half of the eighteenth century and developed rapidly, spreading to Europe and beyond. It resulted, among other things, in a massive increase in the use of non-renewable resources (such as fossil fuels) as well as greater use of renewable but depletable natural resources.

● The information and communication revolution associated mainly with the development of digital technology marks another important stage in economic development. This commenced towards the end of the twentieth century and is on-going. This might be the beginning of the Third Economic Revolution.

In practice many factors (other than changed means of obtaining a livelihood) have contributed to economic growth. These include the development of writing, metallurgy, the use and harnessing of new sources of energy, and improved means of transport. In addition, new forms of industrial and social organization have played an important role in the economic growth process.

The Environment, Control of Nature and Stages of Economic Development

The overall consequence of all these developments has been that humankind has been able to exert increasing control over nature and extract an ever-growing amount of economic provisions from the natural world by transforming it and using it. This has enabled the global human population to increase tremendously and has allowed many people to obtain a high level of consumption of material goods. Thus, God's purported command to newly created humans (according to an extract from Genesis 1:28) that they should 'Be fruitful, fill the earth and conquer it' (Jones, 1966, p. 16) has been to a large extent acted upon. Buddhism, Hinduism and Jainism portray a more inclusive and sympathetic attitude towards nature. Nevertheless, in contemporary societies, the motivation to acquire material goods is very strong, even in countries (such as India, Sri Lanka and Thailand) where these religions predominate. Religious belief seems to have done little to curb materialism in historic times. Consequently, in most contemporary societies (Bhutan might be an exception) natural environments and wildlife are adversely affected by materialism.

Although direct harm to wildlife may be avoided, the indirect aspects of economic growth (these indirect effects include the loss, alteration and reduced variety of natural habitats available to wildlife) designed to increase wealth usually impose a heavy toll on wildlife and natural environments. In general, this toll has increased with the successive 'stages' of economic development.

It should be borne in mind that dividing the process of economic development into discrete stages involves a gross simplification. Globally the process was almost continuous but occurred at different rates of change at different times. It was driven by human migration, new inventions, their diffusion and their adaptation to new environments. Furthermore, crucial points in economic development can be identified in different ways, for example by the type of tools available to humankind, such as whether they were made of wood, stone, copper, bronze or iron, or the source of energy or motive power in use such as human energy, animal power, water and wind and more recent sources of power, for instance those based on fossil fuels. Successive new sources of energy and motive power have made an increasing contribution to economic growth and to the ability of humankind to transform nature.

New forms of social and economic organization have also played an important role in the economic growth process. Palace-based social and economic organizations evolved in some parts of the world following the commencement of agriculture and replaced tribal forms of socio-economic organization. The development of agriculture (particularly the availability of a storable and transportable agricultural food surplus) played a significant role in this evolution. Societies of this type arose in Asia Minor, in ancient Egypt and in the Aegean region, for instance, Minoan and Mycenaean civilizations, in China as well as in the Americas, for instance, the Aztec and Inca civilizations. Depending on how their economic surplus was used, these societies became a springboard for economic growth (Svizzero and Tisdell, 2014a; Tisdell and Svizzero, 2015a; 2015b) and accelerated the intensity and extent of humankind's transformation of nature.

Hunting and Gathering and Environmental Change

Even prior to the adoption of agriculture, the activities of hunters and gatherers exerted a growing impact on nature. This was partly a result of the movement of *Homo sapiens* from Africa to settle eventually in most places on Earth and a consequence of hunting and foraging techniques developed by hunters and gatherers. For example, Australian Aborigines selectively used fire to burn areas of natural vegetation to improve their prospects of successfully hunting some wild species for food. This altered natural landscapes and favoured the survival of particular species (Flannery, 1995; Gammage, 2012). It is also possible that as a result of hunting, they hastened the demise of some ancient wild species. In more recent times, it has been suggested that hunting by Maoris in New Zealand was an important contributor to the extinction of species of moa, large flightless birds (Anon, 2014).

The Development of Agriculture and Environmental Change

Each successive stage of economic development increased humankind's domination of and transformation of nature. The successful growing of crops required substantial changes to the natural environment. Environments needed to be created that were favourable to growing different types of crops and natural ecological competitors had to be controlled. Natural vegetation required removal (by slashing and burning and other means) and preparation of the soil for planting crops was needed. In some cases, irrigation was relied on and manuring of crops was practised in Neolithic times.

Not all early agriculture proved to be sustainable. In Central America (in the Yucatan Peninsula) the clearing of rainforest for agriculture at first supported the Mayan civilization but eventually resulted in depletion of the fertility of this soil and the collapse of this civilization (Haywood, 2010). The 'over-exploitation of fragile rainforest soils brought about the collapse of [Mayan] agriculture' (Haywood, 2010, p. 194). The Mayan population declined dramatically due to starvation caused by this ecological crisis, and kingdoms fought to control the remaining fertile land which also eventually became unproductive due to intensive use (Haywood, 2010, p. 195). Salting of soil from irrigation in the Mesopotamian region also proved to be a problem for sustaining cropping yields. The Smithsonian Institution states that 'as salt built up in the soil, farmers switched to more salt-tolerant grains like barley but the harder they farmed, the less they harvested. After about 2000 years, the once-fertile land of southern Mesopotamia was barren' (Anon, no date).

The Industrial Revolution and Economic, Demographic and Environmental Change

The harnessing of new energy resources, other than human energy, played a very important role in the transformation of nature, as did the development of metallurgy. In addition, increasing possibilities for the exchange of commodities, especially the extension of markets, stimulated economic growth and contributed to changes in economic organization and alterations in social structures. The importance of these factors as contributors to economic growth became particularly apparent during the Industrial Revolution. This revolution is considered to mark a watershed in economic development (Clark, 2007). It ushered in a period of rapid economic growth. However, this should not be attributed entirely to the process of accelerating industrialization. Several factors played a role in this economic take-off. These included European imperial expansion following its Age of Discovery and colonization of overseas territories. These factors resulted in the increased availability of natural resources for industrializing countries, such as the United Kingdom, and provided opportunities for migration of the 'excess' population of some European nations.

The whole process eventually resulted in a rising standard of living for most residents of those countries able to participate in the Industrial Revolution and a reduction in the proportionate number of their poor. The bulk of the population of nations after transiting through the Industrial Revolution eventually had greater levels of income than they were able to obtain after the Agricultural Revolution. In earlier times, the economic benefits of economic growth were restricted to a relatively small dominant class (Svizzero and Tisdell, 2014a; Tisdell and Svizzero, 2015a; 2015b). The bulk of the population remained at subsistence level and this limited population growth. However, depending on how the dominant class used the economic surplus (which they appropriated), it was still possible for significant economic growth to occur accompanied by much environmental change. Nevertheless, these earlier changes pale in magnitude compared to the economic growth, demographic and environmental consequences of the Industrial Revolution.

The Industrial Revolution resulted in a huge increase in global economic production and this increase continues even though its rate of increase may have moderated somewhat. It has resulted in the ever-increasing use of raw materials for productive purposes. Stocks of many natural resources have been reduced. Most natural environments have undergone substantial transformation and considerable loss of biodiversity has occurred as a result of expanding global production. Changes in air, water and soil quality have become of increasing concern as a result of stresses placed on these natural resources by economic production. Many of the problems associated with these changes are of a global nature, for example, global warming as a result of rising levels of GHG in the atmosphere.

Demographic change also occurred as a result of the Industrial Revolution. A consequence has been a massive growth in the level of global population. This is so even though the rate of population increase tapered off in countries that successfully completed the initial industrialization phase of their economic growth. These post-industrialization countries are now highly urbanized and family sizes in these countries have fallen substantially. Nevertheless, these consequences have by no means solved environmental problems associated with global population growth. For example, although the rate of natural population increase is low (even negative) in most higher-income countries, they are beset by immigration from lower-income countries and from regions adversely affected by armed conflict. This is affecting the dynamics of their demographic change.

It should also not be forgotten that the expansion of European settlement into the Americas, Australasia and some other areas has resulted in a tremendous increase in population in these areas and major changes in their natural environments, a process reinforced by the use of new methods of economic production and accelerated economic growth. For example, when the first European (British) settlement began in Australia in 1788 the Aboriginal population was estimated to be 200 000 and relatively stationary. By 2015, Australia's population had reached around 24 million – more than a hundred-fold increase. Immigration has made a major contribution to this increase and is predicted to continue to contribute significantly to further growth in Australia's population. It is not clear when, and if, a ceiling to the level of Australia's population will be reached. What is clear is that major changes to the Australian environment have occurred following its initial British settlement in 1788. The same can be said of many other countries that were colonized following European contact, for instance, the United States and other parts of the Americas. Another example is Mauritius. Prior to the arrival of Europeans, it was uninhabited. Human settlement has entirely changed its natural environment and destroyed many of its wildlife species, of which the dodo, a large flightless bird, is a well-known example.

Volume of Raw Materials Used Continues to Rise Despite the Falling Intensity of their Use

Colin Clark (1957) pointed out that (in recent times) as economic growth proceeds, the proportion of Gross Domestic Product (GDP) accounted for by primary industry declines (for example, agriculture). That accounted for by manufacturing usually rises, and then declines with the tertiary sector (the service sector) growing in relative size. Taking into account the changing economic structure of societies as a result of economic growth as well as the nature of technological progress, it seems likely that the amount of raw materials (primary commodities) consumed in proportion to GDP has risen and then declined in countries experiencing sustained economic growth. The intensity of use of primary commodities, therefore, as a function of economic growth may be of an inverted U-shape. This can be regarded as a Kuznets material–consumption relationship. However, just as in the case of the environmental Kuznets curve (see Tisdell, 2001), this does not imply that the total consumption of primary commodities declines once the Kuznets material–consumption relationship passes its peak.

A fall in the proportion of new materials used in aggregate production may come about in several ways. For example, new techniques often become available which economize on the use of new materials. Secondly, new industries that have low raw material intensities may increase in relative importance. This is true of most service industries and these tend to account for an increasing proportion of GDP as per capita incomes rise.

Whether or not the digital revolution will reduce the intensity of the use of raw materials is unclear but it seems likely. However, even if it does, global consumption of raw materials is likely to continue to increase.

2.3   EHRLICH'S EQUATION, ECONOMIC GROWTH AND ENVIRONMENTAL CONSEQUENCES

Ehrlich's equation (Ehrlich, 1989; Ehrlich et al., 1989) is a useful (but not perfect) device for giving consideration to the historical relationship between the state of the environment and economic growth. It also provides a convenient introduction to considering some conceptual and distributional issues about the environment which have not been given enough attention in the literature. For example, many types of environmental spaces exist, some of which are private goods and others shared goods of different economic types. Considerable inequality exists in the types of environment that individuals can enjoy.

Observations Based on Ehrlich's Equation

Ehrlich (1989) suggested that, as a rough rule, deterioration in the natural environment depends on the level of human population, P, income per head, y, and the nature of technology, T, used for economic production. Other things held constant, the natural environment deteriorates with increases in population levels or with a rise in production (income) per head. The impact of technological change on environmental deterioration depends on whether it is environmentally friendly or not. Other things held constant, new technology can lower or increase the rate of environmental deterioration depending on whether it is 'environmentally beneficial' or not. Ehrlich's relevant equation is:

$$D = f(P, y, T) \tag{2.1}$$

where $\frac{\partial D}{\partial P} > 0$, $\frac{\partial D}{\partial y} > 0$ and the size of $\frac{\partial D}{\partial T}$ depend on the nature of the change in technology.

Even if the change in technology is environmentally favourable, the natural environment may continue to deteriorate because P times y (that is aggregate output, X) continues to rise in a way which negates the potentially beneficial consequences of new technology. In fact, while advances in science and technology have made it possible to reduce the environmental footprint of humankind, this possibility has not been realized. One of the main reasons for this is the desire for ever-increasing economic production

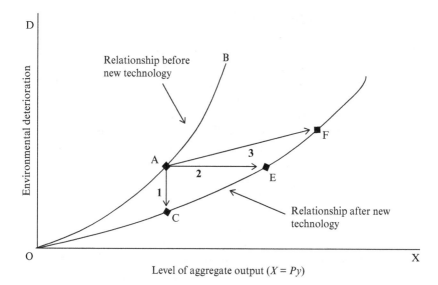

Figure 2.1   The possible implications of Ehrlich's equation as explained in the text

(consumption) using new techniques and this has been accompanied by increases in human population.

Figure 2.1 can be used to illustrate some consequences of Ehrlich's relationship. In this figure, deterioration in the natural environment is shown as a function of aggregate production. Suppose that with existing techniques, the relationship between D and the level of aggregate production is as shown by curve OAB, and that the system is initially stable at point A. Now suppose that new techniques are developed which alter this relationship so that it becomes OEF. Consider three possible changes in the relationship between the state of the environment and aggregate output if the new techniques are adopted. These are:

1. aggregate output is held constant and environmental deterioration is reduced (a movement from A to C);
2. environmental deterioration is held constant and aggregate output is increased (a movement from A to E); and
3. both aggregate output and natural environmental deterioration are allowed to rise (for example, a shift from A to F).

It is often believed that economic development has resulted in the occurrence of a type 3 relationship.

Policy Solutions and Ehrlich's Equation

A number of different policy solutions exist for preventing deterioration in the natural environment given Ehrlich's relationship. These
include encouraging the development and the adoption of environmentally
friendly technology, restricting population growth and limiting increases in
output (consumption) per head. If greater use is made of environmentally
friendly technology, it is possible to have some increase in population or
in income per capita (or both) without environmental deterioration occurring, as is clear from Figure 2.1. Consequently, it is not necessary to have
a steady state economy in which there is zero population growth (ZPG)
and a constant level of per capita income to achieve sustainability of the
natural environment.

Note that in practice it is not only the type of technologies used for
economic activity which affect environmental change but also the composition of production. This is not taken into account in Ehrlich's relationship.
If with the occurrence of economic development, commodities having a
lower intensity to cause environmental change are substituted for others,
this can also contribute to environmental sustainability. Moreover, more
efficient resource allocation can provide an opportunity to reduce the
adverse impacts of economic activity on environmental sustainability.

Although controlling all of the above-mentioned variables can increase
the scope for conserving the natural environment, there is no guarantee
that there will be sufficient political will to adopt such controls. Even if
some of these measures are adopted, the desire for increased economic
growth can result in pro-growth policies that more than offset their favourable environmental consequences. For example, China during its market
reform period had a one-child policy in place until 2016, and to some
extent its market reforms improved its allocation of resource use. However,
China's economic growth and its composition of production were such
that its natural environment deteriorated considerably.

It has been suggested that improvements in the natural environment
are dependent on sufficient economic growth occurring. This has been
typified by the environmental Kuznets curve (see, for example, Selden and
Song, 1994). If the level of population is held constant, it has been argued
that the rate of deterioration in the natural environment will initially
increase with rising income per capita, peak and then decline. Therefore,
it is hypothesized that with continuing economic growth, a deteriorating
economic environment is likely to be replaced by an improving one. Given
this point of view, economic growth is seen as a pathway to eventual environmental improvement. This does not, however, mean that the original
natural environment can be restored, because some environmental changes

are irreversible or may have negative feedback consequences which result in increases in per capita income becoming unsustainable, for example, due to global warming (see Tisdell, 2001). Observe that the environmental Kuznets curve suggests a more optimistic relationship between economic growth and environmental change than does the Ehrlich equation.

2.4 THE VALUE OF MAN-MADE ENVIRONMENTS AND THE RELATIONSHIP BETWEEN ECONOMIC GROWTH AND INEQUALITY IN ACCESS TO ENVIRONMENTAL SPACES AND COMMODITIES

For most humans (probably all), sustaining natural environments is not their primary goal. Most wish to transform natural environments (to some extent) to obtain higher incomes or to create environments that are more agreeable to them than natural environments. However, socio-economic systems can result in greater transformation of natural environments than is desired. As a result, incomes can be lower and less sustainable than they need be and the environments accessed by individuals can be less desirable than they could be. Possible negative effects of loss of natural environments/ecosystems on economic production and incomes have received considerable attention. Here I want to bring attention to environmental aspects of development that have received less attention.

Considering the utility of different possible environments, most humans do not get maximum utility from natural environments alone. If only the utility directly obtained from alternative possible environments is taken into account, some transformations of natural environments provide human beings with increased satisfaction from an aesthetic point of view and often also from the point of view of comfort. For example, shelters are built to increase environmental comfort, landscaping is done for aesthetic effect and architecture can be appealing.

Although it is clear that some natural environments can become more desirable as a result of human modification, there is no guarantee that their socially optimal modification will occur aesthetically or otherwise. For example, methods to create improved environments within homes can result in reduced external air quality. For instance, the burning of coal and other fossil fuels and organic matter can add to smog. The use of air conditioning can have a similar effect depending upon how the electricity used for operating air conditioners is generated. With continuing economic growth, the possibility of individuals isolating or protecting themselves from deteriorating natural environments and unwanted natural environmental effects has increased. This may make them less concerned about some environmental changes, such as global warming, than they should be.

Personal, Private and Shared Environmental Spaces and Resources

Frequently, the attributes supplied by the environment are treated by economists as if they are pure public goods or bads, depending upon whether they have a positive or negative effect on utility. However, this fails to take account of the considerable diversity of environments and differences in the ability of individuals to enjoy liked environments and to avoid or protect themselves from disliked environments.

Some environmental spaces and amenities are effectively private economic goods from which others can be excluded. These include environments in homes and often their immediate surroundings, such as all or parts of their gardens. The same is true of privately owned tourist resorts – the environment in their grounds can be very attractive compared to squalor existing outside. In the latter case, however, the environment within the tourist precinct is a shared good belonging to the club-type category. At least three economic types of shared environment exist. These are:

- club goods;
- quasi-public goods; and
- pure public goods or bads.

Club-types of environment are those that are shared by a particular and restricted group of individuals. Quasi-public environments are those that are open to all but which may be diminished in economic and environmental value if they become congested and quite crowded. These include urban parks and national parks. In the case of pure public goods, their use is non-rivalrous and exclusion is impossible. In the case of a pure public bad, individuals are unable to avoid its adverse effects and these effects are independent of the number of individuals subject to these. In practice, both these pure cases seem to be rare.

The existence value of desired wildlife and some other desirable environmental goods provide examples of pure public goods. Some of these goods are localized ones. For example, good natural air quality and desirable environmental views may only be available in specific locations. Enjoying them may only be possible by those who live in these localities. In the case of environmental bads, even though they can be pervasive (such as poor air quality), it is often possible for individuals to exclude themselves from their negative effects or mitigate them. For example, air conditioning can be used to reduce the negative effects of poor quality of surrounding air. Various other measures are also often available to privately mitigate or avoid possible negative effects on health from unsafe environments. For example, the boiling of water which is contaminated by germs can make it safe to drink.

It is clear from the above that the economic nature of environmental commodities is very diverse – more diverse than is often recognized in the economic literature. Furthermore, the economic nature and relative importance of different environmental commodities alters as economic development occurs. For instance, scientific and technological progress can increase the scope of households to exclude themselves from the ill effects of some types of external environmental conditions.

Environmental Disparity, Income Inequality and Economic Growth

Those individuals with higher incomes are usually able to experience more desirable environments than those who are less well off. Moreover, the desire to enjoy favourable environments normally increases with income. Wealthier persons (compared to those who are less well off) can create more satisfying private environments, can afford to live in neighbour-hoods that are environmentally superior and have greater means to travel to pleasant environmental areas for tourism and recreation. Consequently, income inequality and disparity in the enjoyment of environmental ameni-ties are usually positively associated.

Kuznets (1955) found empirical evidence that in the early stages of eco-nomic growth (relying on industrialization) income inequality increases. Given the above relationships between income and access to environmental amenities, one would also expect growing inequality in enjoyment of envi-ronmental conditions to occur. In the early stages of economic growth, both the incomes and environmental amenities enjoyed by the rich probably increased while the environmental conditions experienced by the masses declined. At this stage of economic development, if the rich constitute only a small proportion of the population, society may show little political inter-est in measures to improve environments shared by many. However, as eco-nomic growth proceeds, this situation can be expected to eventually alter.

As economic growth continues, a higher proportion of the population often obtains higher incomes. Consequently, the aggregate demand for access to shared favourable environmental surroundings tends to increase. This usually occurs at a time when their supply is decreasing or at increas-ing risk of decline. This is likely to result in increasing political pressure for the preservation of these environments by those who are economically better off.

Furthermore, it is relevant to note that following the Industrial Revolution, political representation of the bulk of the population subsequently increased. As a result, the desires of the majority of the population about environmental matters received greater political exposure. Potentially, one would expect this to result in increased political attention being given to members of the community who are not in the highest income bracket. Nevertheless, it still remains true that the richer members of society are able to enjoy superior environments compared to its poorer members. This is because environmental commodities (amenities) consist of diverse types of economic goods, as was explained above.

2.5 SOCIO-ECONOMIC CONSEQUENCES OF SOME PAST MAJOR CHANGES IN NATURAL ENVIRONMENTS, MOSTLY UNRELATED TO HUMAN ACTIVITY

During the existence of *Homo sapiens*, many changes have occurred in natural environments that are independent of human activities. However, several of these natural changes have been exacerbated by human activities, and of course some alterations to natural environments are solely due to human economic activity. Nevertheless, many major past changes in global environments cannot be attributed to human activity. These include major variations in global warming and cooling. Sustained periods of global warming have been associated with rising sea levels whereas substantial periods of cooling (in which a greater volume of water is locked up in ice) have been associated with declining sea levels.

Figure 2.2 presents a stylized representation of the main features of historic sea levels in the last 140 000 years. Sea levels were approximately 140 metres below current levels 140 000 years ago and subsequently increased to slightly above current levels about 120 000 years Before Present (BP). They then declined with fluctuations (not shown in the graph) falling abruptly around 35 000 years BP to reach a minimum level of 140 metres below current levels by 20 000 years ago. At this time, the last glacial maximum reached its maximum extent. Since then, sea levels have risen to reach current levels as global temperatures have become higher and global ice cover has declined. All of these changes have impacted on human settlements. In periods of rising sea levels, human settlements inundated by the sea have had to be abandoned but were resettled as sea levels fell. Furthermore, associated global climate changes would have undoubtedly triggered migrations. These were facilitated in some cases by the existence of 'land bridges'.

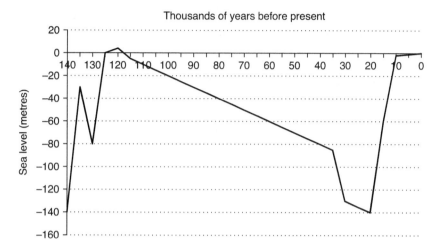

Source: Based on a CSIRO image accessed 28 January 2016 at http://www.azimuthproject. org/azimuth/show/Sea+level+rise.

Figure 2.2 Stylized pattern of sea levels during the past 140 000 years

The last 10 000 years or so has been a relatively warm period globally due primarily to natural causes. Nevertheless, significant cold periods have occurred during this period and have affected human populations (Cunliffe, 2015). However, it is only since about 1750 that human influences have begun to have a significant effect on global temperatures, pushing them beyond the highest average levels estimated for the Holocene era, and well above those recorded when industrialization first began.

Major changes occurred in regional environmental conditions during the Holocene era. It has been suggested that, in early millennia in this period, temperate forests in China and Japan extended northward, and due to higher rainfall current desert areas of Central Asia were covered in forest (Anon, 2016). Furthermore, Western Africa and the Sahara were wet 16 000 to 6000 years ago. As Wikipedia explains, 'during this period, the "Green Sahara" was dotted with numerous lakes containing typical lake crocodile and hippopotamus fauna' (Anon, 2016). Both Central Asia and West Africa (including the Sahara) have become drier in more recent times. Natural changes in global and regional climatic conditions are believed to be influenced in part by long-term periodic alterations of the Earth's orbit and tilt as it moves around the sun (Anon, 2016; Stacey and Hodgkinson, 2013). Volcanic activity is an additional influence. High levels of volcanic activity can change climatic conditions for a considerable amount of time over a large geographical area. Changes in sea temperatures and variations in major marine currents can also be important in altering regional weather patterns.

With economic growth and development, human activities have contributed significantly to regional environmental change and more recently to global environmental change. Human land use in arid and semi-arid regions of North Africa, such as the Sahel, and in parts of China has hastened desertification. In many cases, cropping and grazing pressures substantially reduced vegetative cover, making the land prone to soil erosion. Just how much of this environmental change should be attributed to natural causes and to human activities is unclear. What, however, is apparent is that in the early Neolithic period, human influences on the natural environment were minimal but became increasingly important with the passage of time. For instance, they became increasingly important in North Africa particularly following the expansion of the Roman Empire. A similar situation seems to have emerged in areas of China experiencing increasing aridity.

Human Response to Changing Environmental Conditions

Several types of human response to regional changes in natural environmental conditions are possible. One strategy is to engage in emigration to areas which have more favourable environmental conditions. There is significant historical evidence of such movements in relation to regional climate change (Cunliffe, 2015). However, environmental immigrants are likely to come into conflict with those already residing in more favourable environmental regions if these areas are already occupied.

Another strategy is to change the nature of economic activities and methods of production in response to regional climate change. For example, cropping might be increasingly replaced by grazing. In such a case, however, the land is liable to support a lower level of population because economic production is reduced. Therefore, some community members in areas experiencing environmental stress might migrate or the lower levels of economic output may reduce birth rates and increase death rates, thereby lowering the level of the resident population.

Migration as a Response to Environmental Change

One should also consider the likelihood that inertia is present in human adjustment to regional climate change. Those in areas suffering environmental deterioration may be reluctant to change techniques because of the strength of habitat and uncertainty about the progress of climate change.

They may prefer to emigrate to a natural environment similar to their pre-existing one if environmental conditions deteriorate enough in the region settled by them. This can result in few, if any, of the original population remaining in the affected region. This can then provide an opening for other groups possessing different and more suitable methods of land use to occupy the lands of those who have not adapted to local climate change. It has been suggested, for example, that Arab pastoralists were motivated to settle in parts of north-east Africa once it became less suitable for cultivation, or less able to support hunting and gathering (Anon, 2016; Cunliffe, 2015). Environmental change may well have been an important influence on the diffusion of pastoralism (Cunliffe, 2015).

2.6 THE OPTIMALITY OF INCREASED EXPLOITATION OF NATURAL RESOURCES (THEIR REDUCED CONSERVATION) WHEN THEIR LOSS IS ACCELERATED DUE TO UNCONTROLLED FACTORS

Given that regional land-use practices influence regional environmental deterioration, and that some of this deterioration occurs independently of these practices, what difference should this make to the optimal economic choice of land-use practices? The answer is that the net economic benefit from accelerating the (current) use of regional natural resources (reducing their conservation for future use) increases as the rate of independent loss of these resources due to uncontrolled environmental conditions accelerates. In other words, the marginal user costs of conserving natural resources declines.

This is illustrated in Figure 2.3. There line ABD represents the 'current' marginal regional economic benefit from utilizing natural resources with increasing intensity. The inverse of this is the reduced conservation of these resources. Lines OBF and OCG represent two alternative possibilities for the marginal user costs of 'currently' utilizing these resources. They indicate marginal future economic benefits forgone by current use. Line OBF represents a situation in which uncontrolled environmental change allows these resources to be conserved in greater volume and for a longer period of time than would be so if the marginal user-cost schedule is shown by OCG. If the former situation prevails, the net regional benefit of natural resource use is maximized for an intensity of use of x_1. In the latter situation, however, it is regionally optimal (from a strictly economic point of view) to increase the rate of natural resource use from x_1 to x_2, and reduce the conservation of these resources.

The above theory, however, does not imply that policies for the

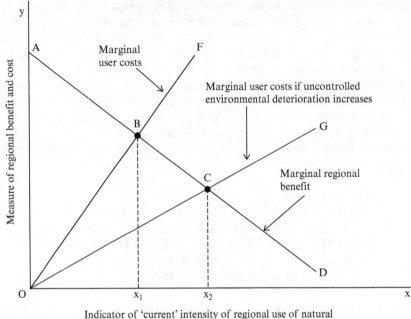

Figure 2.3   *An illustration of the proposition that an acceleration in*
the regional loss of natural resources (due to uncontrolled
environmental causes) increases the economic incentive
to accelerate their use and reduces the economic worth of
conserving them

economically optimal use (conservation) of natural resources will be
adopted. For example, excessive current use (insufficient conservation)
of natural resources may still occur. Nevertheless, in many situations,
account needs to be taken of uncontrolled changes in natural resource
availability. These can result in reduced amount of future natural resource
availability in some regions, irrespective of human actions. On the other
hand, in some cases, these independent natural forces can result in greater
future abundance of valuable natural resources. This could also reduce the
actual incentive to conserve current available natural resources in regions
expected to have greater abundance of valuable natural resources in the
future, because this can also reduce marginal user cost.

2.7 CONCLUDING COMMENTS

Very general relationships between the historical development of economic activity and environmental change (and vice versa) have been discussed in this chapter. Even in the earliest stages of economic development, human beings began to alter natural environments and this process gathered momentum as global economic growth and development proceeded.

Ehrlich's equation for analysing the impact of economic growth on the state of the natural environment was considered. It is commonly used to convey the message that all other things being held constant, economic growth results in the natural environment deteriorating. However, this can be offset by the adoption of technologies that are environmentally friendly. Nevertheless, as shown, this is not sufficient to ensure that deterioration in the natural environment is avoided. Policy prescriptions suggested by this equation were discussed. It is unlikely that sufficient political will exists to adopt several of the proposed solutions, such as limiting personal consumption and population growth. An alternative view is portrayed by the environmental Kuznets curve, namely that with sufficient economic growth it is possible to have higher per capita incomes, as well as to improve the state of the natural environment. This is a politically attractive message. However, it is doubtful if this relationship can be generalized. Both at regional levels and at the global level, increasingly economic growth can become unsustainable due to environmental changes attributable to economic growth itself. An example given in this chapter is the collapse of the Mayan civilization.

Humans have had to contend with major uncontrolled natural environmental changes in the past. Some of these were specified and ways in which humans have responded to these were considered.

Particular attention was given to the varied nature of environments classified as different types of economic commodities and to differences in the ability of individuals to enjoy these environments and shield themselves from environmental bads. It was argued that, in general, those on higher incomes are able to enjoy superior environments. It was hypothesized that inequality in the enjoyment of environments probably varies in a similar manner to that indicated by the Kuznets income inequality curve. Factors which are likely to influence political concern about changes in the natural environment as economic growth proceeds were identified.

Due to uncontrolled factors, it is impossible to avoid environmental change in most parts of the world, particularly over long periods of time. In other words, sometimes loss of natural resources due to uncontrolled (natural) environmental change cannot be avoided. The economic implications of the loss of natural resources due to uncontrolled environmental change were considered. In such circumstances, it may be impossible to maintain regional income levels. Discussions of sustainable development (the subject of the next chapter) should include consideration of this possibility.

REFERENCES

Anon (2014), 'Moa', *Wikipedia*, accessed 10 April 2014 at http://en.wikipedia.org/wiki/Moa.
Anon (2016), 'Holocene climatic optimum', *Wikipedia*, accessed 21 January 2016 at https://en.wikipedia.org/wiki/Holocene_climatic_optimum.
Anon (no date), 'In ancient Mesopotamia organized farming fed the first citizens 4,000 years ago', Forces of Change, Smithsonian Museum of Natural History, accessed 1 February 2016 at http://forces.si.edu/soils/02_08_01.html.
Brewer, A. (2008), 'Adam Smith's stages of history', Discussion Paper No. 08/601, Bristol: University of Bristol.
Childe, V.G. (1936 [1966]), *Man Makes Himself*, London: Collins.
Clark, C.W. (1957), *The Conditions of Economic Progress*, 3rd edn, London: Macmillan.
Clark, G. (2007), *Farewell to Alms: A Brief History of the World*, Princeton, NJ: Princeton University Press.
Cunliffe, B. (2015), *By Steppe, Desert and Ocean*, Oxford and New York: Oxford University Press.
Ehrlich, P.R. (1989), 'Facing the habitability crisis', *BioScience*, **39**, 480–82.
Ehrlich, P.R., G.C. Daily, A.H. Ehrlich, P. Matson and P. Vitousek (1989), *Global Change and Carrying Capacity: Implications for Life on Earth*, Stanford, CA: Stanford Institute for Population and Resource Studies.
Flannery, T. (1995), *The Future Eaters: An Ecological History of the Australian Lands and People*, Sydney: Grove Press.
Gammage, B. (2012), *The Biggest Estate on Earth: How Aborigines Made Australia*, Sydney: Allen and Unwin.
Haywood, J. (2010), *The Ancient World*, London: Quercus.
Jones, A. (ed.) (1966), *The Jerusalem Bible*, London: Darton, Longman and Todd.
Kuznets, S. (1955), 'Economic growth and income inequality', *American Economic Review*, **45**, 1–28.
Malthus, T.R. (1798), *An Essay on the Principle of Population as it Affects the Future Improvement of Society*, (Reprint edn: 1976, Norton, New York) London: J. Johnson.
Markoe, G.E. (2000), *Phoenicians*, London: The British Museum Press.
Meek, R., D. Raphael and P. Stein (eds) (1978), *Adam Smith: Lectures on Jurisprudence*, Oxford: Clarendon Press.
Renfrew, C. (2007), *The Making of the Human Mind*, London: Weidenfeld and Nicolson.
Selden, T.M. and D. Song (1994), 'Environmental quality and development: is there a Kuznets curve for air pollution emissions?', *Journal of Environmental Economics and Management*, **27**(2), 147–62.
Stacey, F.D. and J.H. Hodgkinson (2013), *The Earth is a Cradle for Life: The*

Origin, Evolution and Future of the Environment, Hackensack, NJ: World Scientific.

Svizzero, S. and C.A. Tisdell (2014a), 'Inequality and wealth in ancient history: Malthus' theory reconsidered', *Economics & Sociology*, **7**(3), 223–40.

Svizzero, S. and C.A. Tisdell (2014b), 'Theories about the commencement of agriculture in prehistoric societies: a critical evaluation', *Rivista di Storia Economica*, **30**(3), 255–80.

Svizzero, S. and C.A. Tisdell (2015a), 'The persistence of hunting and gathering economies', *Social Evolution & History*, **15**(2), 3–25.

Svizzero, S. and C.A. Tisdell (2015b), 'Economic management of Minoan and Mycenaean states and their development', *Rivista di Storia Economica*, **31**(3), 341–93.

Svizzero, S. and C.A. Tisdell (2016), 'Economic evolution, diversity of societies and stages of economic development: a critique of theories applied to hunters and gatherers and their successors', *Cogent Economics & Finance*, **4**(1), 1161322.

Tisdell, C.A. (2001), 'Globalisation and sustainability: environmental Kuznets curve and the WTO', *Ecological Economics*, **39**(2), 185–96.

Tisdell, C.A. and S. Svizzero (2015a), 'The Malthusian Trap and the development in pre-industrial societies: a view differing from the standard one', *Social Economics, Policy and Development*, Working Paper No. 59, Brisbane: School of Economics, The University of Queensland.

Tisdell, C.A. and S. Svizzero (2015b), 'Rent extraction, population growth and economic development: development despite Malthus' theory and precursors to the Industrial Revolution', *Economic Theory, Applications and Issues*, Working Paper No. 73, Brisbane: School of Economics, The University of Queensland.

3. Sustainable (economic) development: what is it? Is it desirable? Can it be achieved and if so, how?

3.1 INTRODUCTION

The concept of sustainable development first achieved international prominence in 1972 as a result of the first United Nations Conference on Environment and Development (UNCED) held in Stockholm. Since then, it has been a key element in international dialogue about development policies and has been given much scholarly attention. However, the subjects of what is sustainable development, whether it is desirable, and the possibility of achieving it (and how) are by no means settled. Furthermore, since 1972, international politics in relation to this matter appears to have altered. At the UNCED Conference held in 1972, the group of Non-Aligned Nations (led by India) expressed concern that the economic growth of higher-income nations might threaten their economic prospects as a result of natural resource depletion and pollution. Since then, high-income nations have become increasingly concerned about the environmental consequences of economic growth in lower-income countries which are experiencing substantial economic growth, such as India and China. At the 2015 Paris Conference on Global Climate Change, the Indian delegation took the view that only by fostering economic growth can India eventually reduce its adverse environmental footprint. This view relied implicitly on the inverted-U relationship between the state of the environment and economic growth projected by the environmental Kuznets curve (see the previous chapter). So to some extent, the political tables on this issue have turned. However, the general issues involved remain of global concern.

This chapter provides a broad (rather than a detailed) coverage of issues involving the concept of sustainable development and its applications, particularly sustainable economic development. A more detailed coverage of several of these issues can be found for example, in Tisdell (2015, Ch. 4). This chapter begins by considering different criteria for sustainable economic development and then considers the desirability of the different economic growth patterns which they rank, and whether the actual concerns of current generations for future generations are likely to accord with those suggested preference rankings. Attention then turns to considering and assessing conditions proposed (weak and strong) for achieving sustainable economic development. The trilogy of conditions (three pillars) claimed in the relevant literature to be required for achieving sustainable development are then critically examined. It cannot be assumed that sustainable economic development is always possible. In several circumstances, options for maintaining income levels and human well-being are limited or non-existent. Attention is given to the policy consequences of this. Achieving environmental/ecological sustainability is sometimes portrayed as being necessary to alleviate poverty and/or to curb an increase in the incidence of poverty. This aspect is considered prior to a concluding discussion of the views raised in this chapter.

3.2 DIFFERENT CRITERIA FOR THE OCCURRENCE OF SUSTAINABLE ECONOMIC DEVELOPMENT

Several different sets of criteria for the occurrence of sustainable economic development have been proposed. They most frequently include the following:

1. it is development that does not reduce the economic options of future generations;
2. it is development that results in the income of future generations not being less than that of current generations;
3. it is development ensuring that the income of each future generation is not less than that of its predecessor.[1]

The first condition was suggested in the report of the World Commission on Environment and Development (WCED, 1987). The second is to be found in some economic textbooks (for example, Tietenberg, 2003) and the third was proposed by John Rawls (1971). Criterion 2 is a weaker one than Criterion 3 because it does not require the income of each succeeding generation to be as large or greater than that of its predecessor. If there is more than one development path that satisfies these criteria, any path which is Paretian dominant is to be preferred. In the case of Criterion 3, this can be a development path for which the incomes of some future generations are lower than their predecessors, as Rawls points out.

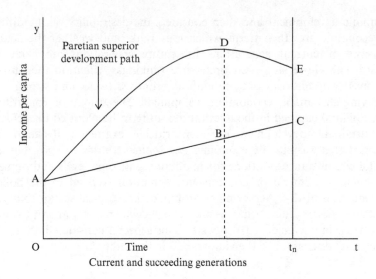

*Figure 3.1   An illustration that some development paths satisfying
            particular criteria for sustainable economic development are
            not the most socially desirable ones, as explained in the text*

This is illustrated in Figure 3.1. Two alternative development paths (ABC
and ADE) are shown. Both satisfy Criterion 2, given that the income per
capita of the current generation(s) is OA. Although Criterion 3 is not satisfied by development path ADE, it is Paretian superior to path ABC, and
therefore is socially preferred given Rawls' Paretian qualification to the
desirability of sustainable economic development.

**Further Qualifications to the Social Desirability of Sustainable Income
Paths**

Rawls bases his requirements for sustainable development on the assumption that these are the conditions all persons (yet to be born) would agree to
prior to their birth if they were able to confer at this time and did not know
when they would be born. This is a completely hypothetical situation. He
assumes that unborn individuals will be extremely risk-averse. However,
the possibility cannot be ruled out that individuals would wish to adopt
the development option that maximizes their expected income, provided
their income does not fall below a minimum satisfactory level. This is an
application of Tisdell's variation of Roy's safety first principle (Roy, 1952;
Tisdell, 1962) and has previously been applied to the choice of alternative
development paths (Tisdell, 2011; 2015, Ch. 4).

In discussing sustainable development, considerable emphasis has been placed on the welfare of future generations. This can result in insufficient attention being given to the future economic welfare of current generations. Given the rapidity of economic and environmental change and the longer length of life being achieved by many belonging to current generations, lack of economic and environmental sustainability can occur within their lifetime. Impoverishment due to environmental changes (sometimes as a result of economic factors) can occur within the lifetime of individuals, and this needs to be factored into economic choices.

How Much Sacrifice Should be Made by Current Generations to Benefit Future Ones?

Discussions of desirable patterns of sustainable development often do not make it clear how much economic sacrifice current generations should make to contribute to the economic welfare of future generations. This matter can be considered heuristically by reference to Figure 3.2.

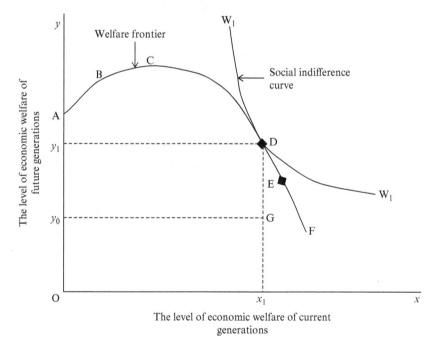

Figure 3.2 Hypothetical trade-off possibilities between the level of economic welfare of current generations and that of future generations used to illustrate problems discussed in the text

In this figure, the curve ABCDF shows hypothetically the trade-off frontier between the economic welfare of current generations and that of
future generations. It allows for the possibility that some increase in current
economic welfare can add to the economic welfare of future generations
as shown by the segment ABC. A social preference function is supposed
to exist which satisfies the usual properties associated with preference
relationships. This might be a social welfare ordering by current generations of the type suggested by Bergson (1938). The social welfare indifference curve, W_1W_1, is the highest attainable one and point D corresponds
to the optimal attainable combination of current and future economic
welfare. Apart from the fact that considerable uncertainty exists about
the nature of the benefit-possibility frontier, several other difficulties exist
which can result in failure in achieving social optimality, given this model.

   The first thing to note is that future decisions about resource use
depend only to a limited extent on current decisions. For example, after
their death, today's decision-makers will have little or no influence on
future decision-makers, and this influence is likely to diminish further into
the future. Consequently, if today's decision-makers choose economic
benefit, x_1, there is no guarantee that future decision-makers will adopt
environmental and other policies which result in y_1 being realized. They,
for example, may act in a way which results in a future economic benefit
of only y_0, an inferior outcome according to the perception of current
decision-makers. Furthermore, future decision-makers may have a different social preference function to current decision-makers. Daly (1999) has
brought attention to the lack of control of each generation over the decisions of its successors.

Actual Concern of Current Generations for the Welfare of Future Generations and for Other Persons

While considerable attention has been given to normative aspects of
intergenerational equity, much less attention has been given to determining empirically the concern of current generations for the welfare of
future ones. Pearce (1998) has suggested that the concern of individuals for the welfare of future generations only extends to a limited extent
(or at all) beyond concern for the well-being of their children and their
grandchildren. This hypothesis is plausible. Moreover, biases are likely
to be present in altruism in relation both to existing and future generations. For example, identification with a particular social group may result
in individuals caring much more about the welfare of this group than other
groups. Pearce's observation indicates that it would be worthwhile empirically determining relevant coefficients of concern.

Another aspect of interest is the ability of individuals to cater (as a result of their own individual decisions) for the welfare of future generations. In many cases, collective action is required to sustain the welfare of future generations, for example, in relation to the provision and conservation of public goods (Daly, 1999, pp. 115–16; Margolis, 1957).

3.3   PROPOSED CONDITIONS (WEAK AND STRONG) FOR ACHIEVING SUSTAINABLE DEVELOPMENT

As pointed out by Pearce (1993, Ch. 2), a wide range of views exist about the general conditions which need to be satisfied if sustainable economic development is to occur. They range from very strong to very weak conditions about the need to conserve natural environments and resources. The chart shown in Figure 3.3 provides a general overview of the spectrum of these views.

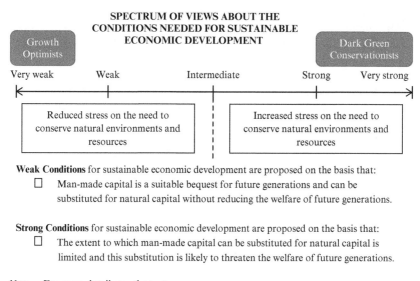

Weak Conditions for sustainable economic development are proposed on the basis that:
☐   Man-made capital is a suitable bequest for future generations and can be substituted for natural capital without reducing the welfare of future generations.

Strong Conditions for sustainable economic development are proposed on the basis that:
☐   The extent to which man-made capital can be substituted for natural capital is limited and this substitution is likely to threaten the welfare of future generations.

Note: For more details see the text.

Figure 3.3   An impression of the range of different views about the need to conserve natural environments and resources in order to achieve sustainable economic development

Reasons for Proposing Weak or Strong Conditions for Sustainable Economic Development

Those who believe that little or no special effort is required to conserve natural environments and resources in order to achieve sustainable economic development usually do so on the basis of one or more of the following propositions:

- Man-made capital is a suitable (often a superior) substitute for natural resource capital.
- The market system on the whole optimizes this substitution and promotes economic sustainability.
- Scientific and technological progress will provide humankind with the ability to more than compensate for any negative effects resulting from the loss of natural capital.

Therefore, little or no public intervention is required to conserve natural capital. The economic analysis of Hartwick (1977) lends support to the hypotheses listed in the first two bullet points above. However, his theory is not robust or general enough to dismiss all objections to the weak approach to sustainable economic development.

Opponents of a weak approach to sustainable development (those favouring strong measures to conserve natural capital) criticize it on the following grounds:

- There are significant limits to the extent to which human well-being can be sustained by substituting man-made capital for natural capital.
- The market system is flawed as a mechanism for ensuring sustainable economic development. Even with market reforms (which can reduce its shortcomings), there are substantial limitations on the ability of the market system to promote sustainable economic development.
- Scientific and technological progress is uncertain and technological 'fixes' can be impossible or prohibitively expensive for some emerging environmental or natural resource scarcity problems likely to threaten sustainable economic development.

Does the Substitution of Man-made Capital for Natural Capital Imperil the Welfare of Future Generations?

The extent to which man-made capital can be substituted for natural capital without imperilling the welfare of future generations probably depends on the scarcity of the stock of natural capital. As natural environments and resources become scarcer (as a result of the cumulative effect of economic growth on these), the difficulty of achieving sustainable economic development by substituting man-made capital for natural capital probably increases (Tisdell, 2005, Ch. 11). In the early stages of economic development, loss of natural environments and resources is necessary to obtain economic development (Tisdell, 2005, Ch. 11). Consequently, the implication of the loss of natural resources for the well-being of humanity varies with the passage of time.

Also note that individual attitudes to whether strong or weak conditions for sustainable development are desirable depend on how much weight people place on conserving nature as an end in itself. Those who place considerable emphasis on this (that is, who hold ecocentric values rather than anthropocentric ones) can be expected to favour strong conditions for sustainable economic development. In fact, they are likely to be prepared to forgo some human well-being for the sake of conserving nature, and are prepared to do so the more strongly they want to conserve nature.

3.4 THE THREE PILLARS REQUIRED FOR SUSTAINABLE DEVELOPMENT: THE TRILOGY CONCEPT

There are claims in the literature (Barbier, 1987; WCED, 1987) that it is necessary for development strategies to be simultaneously, socially, environmentally and economically sustainable if sustainable development is to occur. While this approach highlights the need to consider simultaneously social, economic and environmental factors in assessing strategies for sustainable development, this trilogy concept may be of limited worth from an operational or pragmatic point of view.

Operational Limitations of the Trilogy Approach

This approach is of limited operational value for the following reasons:

- Specifications of each of its sustainability objectives are often left open or are unclear.
- Even if these specifications are adequate, no development strategy may be available to meet the three objectives simultaneously. For example, the type of situation illustrated in Figure 3.4 is likely to be rare or may not occur at all. This Venn diagram identifies development strategies which satisfy each of the specific objectives or

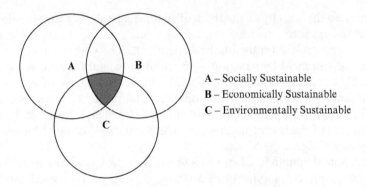

Note: Only those strategies in the shaded area simultaneously satisfy the three requirements for sustainable development. However, as explained in the text, this approach is of limited operational value because the three sets are unlikely to intersect.

Figure 3.4   A Venn diagram illustrating the three-pillar approach to choosing sustainable development strategies

pillars. Only those in the shaded overlapping area satisfy all of these objectives.

● Given that the three criteria for sustainable development are likely to be impossible to satisfy, additional specification is needed to determine the optimality of alternative available development strategies, for example, the extent to which social sustainability and environmental sustainability should be forgone to achieve economic growth. Mathematically, the optimality of choices is under-determined.

Satisfying each of these pillars is unlikely to be an imperative. For example, communities may be willing to forgo some features of social sustainability or environmental sustainability to achieve economic growth or vice versa. In most cases, trade-offs in relation to economic, social and environmental options seem to be unavoidable.

3.5   DIFFERENT INDICATORS OF SOCIAL SUSTAINABILITY

Sustaining Community

The idea has been expressed that social sustainability requires the sustainability of community, for example, local communities, ethnic and minority communities (Douglass, 1984). However, as a general objective, its desirability is difficult to determine. Some local communities may be willing to make economic sacrifices to maintain their cultural identity, social bonding and shared amenities. Despite this, economic changes may make this difficult or impossible for them to achieve this objective. Economic and environmental change can, for instance, result in the 'unstoppable' depopulation of some regional areas and towns with adverse consequences for the availability of their shared facilities and opportunities for social interaction.

Maintaining Social Cohesion

Social cohesion is sometimes claimed to be another attribute of social sustainability worth maintaining. Sometimes, however, social cohesion can only be maintained by force. Opinions are likely to be divided about the extent to which the use of force for this purpose is justified and the circumstances in which this is so.

Maintaining or Increasing the Stock of Social Capital

An additional possible social sustainability consideration is whether the stock of social capital is maintained, increased or decreased by economic and environmental change, or vice versa. Many different items can be considered to be a part of the stock of personal social capital but there is a lack of clarity about their relative importance and how best to measure these and their contributions to the stock of social capital (Antoci et al., 2012; Fukuyama, 2001; Salahuddin et al., 2016, pp. 46–7).

Trust is often identified as a very important component of social capital (Westlund and Adam, 2010). This seems to be because lack of trust reduces the prospect for economic and social exchange and causes psychological stress. Therefore, distrust results in human well-being being lower than it could be. The ability to develop social networks in order to facilitate exchange to tap economic opportunities and obtain social support in times of need is also a personal social asset.

On a different plane, institutional structures (such as codification of the law, the rule of law and some social conventions and morals) can be regarded as important components of collective social capital. Whether or not sustaining these existing structures is desirable depends on their nature and other considerations. In a society in which dishonesty, cheating, corruption and lack of respect for the property of others prevail, economic and environmental conditions are usually worse than they need be. Good governance adds to the stock of collective social capital. However, consensus does not prevail about what constitutes good governance.

3.6 THE APPLICABILITY OF THE TRILOGY OF SUSTAINABILITY CONCEPTS TO THE ASSESSMENT OF DEVELOPMENT PROJECTS

While the three-pillar requirements for determining policies for sustainable development have significant operational limitations, with modifications they have greater relevance to the assessment of development projects. If development projects are to succeed, it is important that they should be socially acceptable, environmentally suitable and economically viable. Failure to meet one or more of these requirements can result in their failure. Consequently, assessment of potential development projects needs to take these three factors into account.

Debates about sustainable development have made an important contribution by highlighting the need for a more holistic approach to assessing development projects than was common prior to the commencement of this debate. This is one important contribution of this debate. Sometimes it is also possible to identify development policies which achieve these three criteria. As I have argued elsewhere (Tisdell, 2014, Ch. 9), public policies for the conservation of saltwater crocodiles in the Northern Territory of Australia have proven to be environmentally sustainable, and to be economically and socially viable.

3.7 POVERTY AND SUSTAINABILITY ISSUES

Only a few aspects of poverty and sustainability can be considered here. Attention will be given first to the *environmental poverty trap* resulting from excessive use of natural resources and to the likely futility of policy advice recommending that the way to escape from this trap is to reduce resource exploitation. The second matter discussed is how sustaining community can trap some social groups (such as ethnic minority groups) into a situation of persistent poverty. This discussion highlights several of the difficulties involved in ensuring that all groups in society are able to achieve sustainable development.

The Environmental Poverty Trap

In communities where incomes are very low, savings are also low and so, consequently, is capital accumulation. Hence, little or no economic growth occurs in these communities and (as pointed out in the development economics literature) they are caught in a poverty trap. It is, however, necessary to consider particular reasons why some communities are unable to escape from a low level of economic development. One possibility is that over-exploitation of renewable natural resources occurs and results in this outcome.

This over-exploitation may arise as a result of social changes within a community, or it may be due to a reduction in the availability of natural resources caused by external factors, or both. Consider the first possibility initially.

As a result of general socio-economic change within the wider society in which a minority community is embedded, communal rules governing the use of shared renewable natural resources may be relaxed or may become less enforceable. Consequently, an open-access situation can develop in resource use. At first, this is likely to result in a rise in per capita incomes in the focal community but these higher levels of income are unlikely to be sustainable. Incomes will prove difficult to maintain as the stock of natural capital is reduced, and in the end may fall to subsistence level and result in persistent poverty for the community concerned. This assumes that the focal community is unable to use their initial increased level of income (as a result of reducing their natural resource stock) to accumulate man-made capital sufficient to compensate them (or more than compensate them) for their loss of natural capital. In addition, it supposes that community members are unable to migrate to different geographical areas and engage in economic activities which would enable them to escape from this environmental trap.

In such circumstances, advising a community which is already in an environmental poverty trap to reduce natural resource use in order to increase incomes is of no practical value (see Tisdell, 2009). This is because of the dynamics involved. If incomes in the focal community are already at subsistence level, reducing the rate of utilization of renewable natural resources (to enable recovery of their stock or improve their quality) will result initially in the per capita income of such a community falling below subsistence level. Unless external economic support is provided to this community in the early stages of this adjustment process (or unless members of the affected community are able to migrate elsewhere to obtain an adequate livelihood), adoption of the proposed environmental policy for poverty alleviation is socially unacceptable. This is because the policy advice is based on comparative statics and does not take into account the dynamics of adjustment.

Examples of the Environmental Poverty Trap

The supply of renewable natural resources in the Okavango Delta of Botswana depends on the seasonal flooding of this delta. However, the extent and length of this flooding has been declining due to the reduced supply of water from streams feeding into it, partly due to use of water upstream for irrigation. As a result, groups depending on renewable resources in this delta have been forced to harvest these resources more intensively to maintain their already low levels of income. This makes it increasingly difficult to maintain their incomes. As a result they appear to be trapped.

In north-east India, as in other parts of Asia, some communities still rely on shifting agriculture for their livelihood. This is so for some Mizo communities, present in Mizoram in north-east India. For various reasons, these communities have reduced the length of time for which they fallow land before using it again for agriculture. Consequently, the fertility of soil used for agriculture is declining and crop yields are falling. The difficulty of maintaining income levels is increasing. Although lengthening the fallow period could potentially raise overall yields, to do so requires land in use to be initially cultivated for an increased period of time, or the area under cultivation at any one time to be reduced. This will depress initial yields and reduce already low levels of income. Therefore, many of these communities are caught in an environmental trap.

Furthermore, many communities comprised of artisanal fisheries are caught in this type of trap. Their fish stocks are over-exploited. However, they cannot bear the initial burden of having their already low levels of income reduced by adopting conservation techniques which will eventually increase the level of their fish stocks and thereby eventually raise their incomes. In addition, because there are likely to be considerable economic and social barriers to the migration of artisanal fisheries, they are locked into an environmental poverty trap. An example is the situation of artisanal fishers in Thailand who use small meshed nets to catch fish. Larger-sized mesh would result in an increase in fish stocks and would eventually raise the level of the sustainable catch. However, switching to nets with larger mesh is not practical because initially it lowers the catch and depresses already low incomes in these communities. The prospects of the villagers involved migrating to avoid this consequence are often slim, and external aid to compensate for their initial income loss is unlikely to be forthcoming.

Social Factors Contributing to Poverty Lock-in and Economic Inertia

Communal ties can be a barrier to the economic sustainability and development of some communities. Even if there is scope for community members to migrate to escape from poverty, they may fail to do so because of social bonding and support available within their own community. Outside their own community they may face social discrimination and lack of social support. In some cases (for instance, if they are an ethnic minority group), they may lose their social identity or have it diminished.

Some minority groups also lack sufficient skill and work experience to facilitate their employment outside their community. This is often coupled with a lack of social contacts outside their community. These factors create major barriers to the migration of some minority groups.

Even when an economic development occurs within a minority area, members of the resident minority group may fail to share in its economic benefit to a significant extent. They may lack relevant skills and experience or face social discrimination in employment. Consequently, the bulk of the economic benefit may be captured by outsiders. Outsiders may have greater relevant know-how, more capital and superior political contacts.

The Case of Australian Aborigines Living in Remote Areas

Australian Aborigines (especially those living in remote areas) on average have lower incomes, a lower life expectancy, a greater rate of unemployment and more health problems than most Australians (see, for example, Tisdell, 2014, Ch. 16). Particularly in remote areas, their comparative level of education is low. This is partly because the supply of educational services is limited and the demand for education is low. Basically, both an educational supply and a demand problem exist. The demand problem seems to exist because actual or perceived economic returns from education obtained by Aborigines living in remote areas are low.

There is also the further problem that many Australian Aborigines find it psychologically difficult to migrate because the cultural identity of each group is closely linked to local landforms and connected to wildlife totems. Often minority ethnic groups living in remote regions in other countries find it difficult to escape from poverty and low levels of income within their communities, as experienced by Australian Aborigines. Furthermore, many of these groups are plagued by property rights problems. Their economic situations are jeopardized by lack of official recognition of their rights in land or natural resources which they have traditionally occupied or used.[2]

3.8 DOES POPULATION MOBILITY INCREASE OR DECREASE THE CONSERVATION OF NATURAL RESOURCES?

Views differ about whether increased population mobility increases or decreases the depletion of natural resources. The following views exist:

- If individuals relying on natural resources for their livelihood are able to migrate to improve their economic lot (or if their offspring can do so), they may have little economic incentive to conserve their natural resources. They may draw on these assets to finance their own migration, or particularly that of their offspring. In the latter case, they may look forward to receiving future remittances from their offspring.
- The above suggests that those who do not have the possibility of migrating will conserve their natural resources because their user costs are high.[3] However, in some cases, their survival depends on the use of their natural resources unsustainably. Even though their user costs are high in this case, so too is their current need to utilize these resources. They may also over-exploit these resources in the expectation that their situation might improve in the future. Furthermore, they may be in a situation where many of the natural resources that they utilize are shared resources and subject to open access or increasing open access. Consequently, if they remain dependent on these resources they have no individual incentive to conserve these.

The possibility was mentioned above that natural resources might be over-exploited to help finance the migration of family members in the expectation of this resulting subsequently in remittances being received in return. This process can include the funding of the education of offspring in anticipation of this assisting them to obtain employment elsewhere that pays well. Although this process may accelerate natural resource depletion initially, once remittances are received the opposite effect is likely to be observed.[4] Furthermore, in the long term, this process is likely to result in families abandoning their traditional location and dependence on natural resources there. This could further reduce pressure on the use of local natural resources. One cannot predict in this case what will happen a priori in the long term. For example, in the long term, in response to migration, subsistence or semi-subsistence farming and natural resource use in a geographical area may be replaced by large commercial farms and enterprises. This can result in the consolidation of rural holdings and privatization of some natural resources. This may or may not be favourable to natural resource conservation. The outcome depends on the particular circumstances.

3.9   CONCLUDING COMMENTS

This chapter has outlined and discussed a wide range of sustainability issues. These include different criteria for the occurrence of sustainable economic development and their limitations. In particular, Rawls' principle of justice and intergenerational equity was questioned. Even if the hypothetical (fictitious) possibility assumed by Rawls of all conferring (on a desirable development path) before being born (not then knowing when they will be born) is accepted, one cannot rule out the possibility that they would opt for a modified form of the safety first rule. Also the question is not considered by Rawls of how his fictitious agreement would be enforced.

Another problem discussed is just how much sacrifice present generations should make for future generations, particularly taking into account the fact that current generations have no control over the decisions of future generations and limited influence on future development paths once they are deceased. Apart from normative views about sustainable economic development, it is also important to investigate empirically how much concern present generations have for the economic welfare of future generations and its nature. Empirical studies of this subject are rare or non-existent. As was pointed out, this concern probably only exists (or has much strength) for a limited number of future generations, and biases probably occur, for example, in favour of one's own offspring, or one's ethnic or social group.

Reasons for organizational failures resulting in desired (and feasible) development paths not being followed are also relevant to debates about sustainable development. A range of views have been expressed about the extent to which natural resources and environments need to be preserved to ensure sustainable economic development. These range from the view that no particular restrictions need to be placed on natural resource use and human-induced environmental change (weak conditions for sustainable development) to the view that considerable intervention is required (strong conditions are relevant). In general, those who favour weak conditions argue that man-made capital is a valuable bequest for future generations and the scope for substituting man-made capital for natural capital and thereby sustaining economic development is considerable. An additional argument for weak conditions often includes the proposition that markets have considerable ability to take account of user costs. Additionally, it is claimed that technological progress can be relied on to counteract the effects of falling stocks of natural capital. Those who favour strong conditions are sceptical about these arguments.

A broader concept than sustainable economic development is sustainable development. It has been suggested that development strategies should be simultaneously socially, environmentally and economically sustainable. A weakness of this approach from an operational point of view is that there are no clear definitions of these 'three pillars' for sustainable development.

Furthermore, if there is no development strategy which satisfies all these criteria, no solution is given to the development problem. Some of the difficulties of specifying these criteria were highlighted by considering several different indicators of social sustainability, such as sustaining community, maintaining social cohesion and social capital. Among other problems, sustaining one of these objectives can have a negative effect on another development criterion, such as the economic sustainability criterion, as well as prospects for economic growth. Particular attention was given to the use of social capital as an indicator of social sustainability. It was found to be a multidimensional concept and difficult, if not impossible, to measure accurately in quantitative terms. However, this does not mean that it is an unimportant consideration in assessing economic development and change.

The trilogy approach to assessing economic development was claimed to be of more operational value for assessing particular development projects (although the criteria may require some modification for this purpose) than in evaluating macro-development. The example was cited of an assessment of the programme to conserve saltwater crocodiles in the Northern Territory of Australia.

The persistence of poverty in some communities and the difficulties they encounter in achieving sustainable economic development were given particular attention. The difficulties that some communities experience in being unable to escape from environmental poverty traps were considered as well as social factors that lock many into poverty or poor socio-economic conditions. The status of Australian Aborigines in remote locations was suggested as an example, but many other minority groups face similar development difficulties. The environmental poverty-trap problem highlighted how relying on comparative static economic models can result in policy advice which cannot be socially (morally) implemented.

Although most contemporary concepts of sustainable economic development (and sustainable development) are of limited operational value and the worth of their normative assumptions can be questioned, discussions of this subject in the last five decades (or so) have been valuable in broadening our assessments of alternative possibilities for economic development. Whether or not these discussions have sufficiently raised awareness about the limitations of economic techniques to predict and measure quantitatively the relative value of development paths in economic change is unclear. Currently, a premium is placed on technical presentations and on quantification by the academic community even though in the social sciences, quantification can be misleading and proxy variables used for measurement can be too narrow to reflect accurately the phenomenon to be measured.[5]

NOTES

1. A different view of sustainable development (expressed by Daly, 1980 and Georgescu-Roegen, 1974) is that it is development enabling the human species to survive for as long as is possible. This view is discussed by Tisdell (1988); see also Tisdell (2005, pp. 251–3).
2. This is a problem experienced, for example, by many Scheduled Tribes in India, for instance forest-dependent Santals.
3. The general hypothesis has been proposed by Klee (1980), for example, that people who depend primarily on (or are locked into) use of their own local ecosystem for their livelihood are more likely to conserve these than those who use global resources. Klee, following Raymond Dasmann, describes the former as 'ecosystem people' and the latter as 'biosphere people'. However, the historical evidence does not support the view that ecosystem people have always conserved existing ecosystems (see Tisdell, 1990, pp. 41–2). Depending on the circumstances, biodiversity loss, for example, may occur in 'isolated' communities as well as in communities participating in the economic globalization process. Examples of the latter types of losses can be found in Tisdell (2015, Ch. 6) and it is known that international trade in endangered wildlife species can (in some circumstances) endanger their survival as is evidenced by the Convention on International Trade in Endangered Species (CITES).
4. Some empirical evidence in support of this hypothesis can be found, for example, in Regmi and Tisdell (2002).
5. An example is the use of the proxy for the trust variable obtained from the World Values Survey (The World Values Survey Association, 2014) which in turn is sometimes used as a measure of social capital (see, for example, Salahuddin et al., 2016). The proxied trust variable is the proportion of surveyed people in a country who answer 'Yes' to the question: 'Generally speaking, would you say that most people can be trusted or that you need to be very careful in dealing with people?'

REFERENCES

Antoci, A., F. Sabatini and M. Sodini (2012), 'See you on Facebook! A framework analyzing the role of computer-mediated interaction in the evolution of social capital', *Journal of Socio-Economics*, **41**(5), 541–7.

Barbier, E.B. (1987), 'The concept of sustainable economic development', *Environmental Conservation*, **14**, 101–10.

Bergson, A. (1938), 'A reformulation of certain aspects of welfare economics', *Quarterly Journal of Economics*, **52**, 310–34.

Daly, H.E. (1980), *Economics, Ecology, Ethics: Essays Towards a Steady-State Economy*, San Francisco, CA: Freeman.

Daly, H.E. (1999), *Ecological Economics and the Ecology of Economics: Essays in Criticism*, Cheltenham, UK and Northampton, MA, USA: Edward Elgar Publishing.

Douglass, G.K. (1984), 'The meanings of agricultural sustainability', in G.K. Douglass (ed.), *Agricultural Sustainability in a Changing World Order*, Boulder, CO: Westview Press, pp. 3–30.

Fukuyama, F. (2001), 'Social capital, civic society and development', *Third World Quarterly*, **22**(1), 7–20.

Georgescu-Roegen, N. (1974), *The Entropy Law and The Economic Process*, Cambridge, MA: Harvard University Press.

Hartwick, J. (1977), 'Intergenerational equity and the investing of assets from exhaustible resources', *American Economic Review*, **66**, 972–4.

Klee, G.A. (1980), *World Systems of Traditional Resource Management*, London: Edward Arnold.

Margolis, J. (1957), 'Secondary benefits, external economies, and justification of public investment', *The Review of Economics and Statistics*, **39**, 284–91, reprinted with corrections in K.J. Arrow and T. Scitovsky (eds) (1969), *Readings in Welfare Economics*, London: George Allen and Unwin.

Pearce, D. (1993), *Blueprint 3: Measuring Sustainable Development*, London: Earthscan.

Pearce, D.W. (1998), *Economics and Environment: Essays on Ecological Economics and Sustainable Development*, Cheltenham, UK and Northampton, MA, USA: Edward Elgar Publishing.

Rawls, J.R. (1971), *A Theory of Justice*, Cambridge, MA: Harvard University Press.

Regmi, G. and C. Tisdell (2002), 'Remitting behaviour of Nepalese rural-to-urban migrants: implications for theory and policy', *Journal of Development Studies*, **38**(3), 76–94.

Roy, A.D. (1952), 'Safety first and the holding of assets', *Econometrica*, **20**, 431–49.

Salahuddin, M., C.A. Tisdell, L. Burton and K. Alam (2016), 'Does internet stimulate the accumulation of social capital? A macro-perspective from Australia', *Economic Analysis and Policy*, **49**, 43–55.

The World Values Survey Association (2014), *The World Values Survey 1981–2014*, accessed 22 February 2016 at http://www.worldvaluessurvey.org.

Tietenberg, T. (2003), *Environmental and Natural Resource Economics*, 6th edn, Boston, MA: Addison Wesley.

Tisdell, C.A. (1962), 'Decision making and the probability of loss', *Australian Economic Papers*, **1**, 109–18.

Tisdell, C.A. (1988), 'Sustainable development: differing perspectives of ecologists and economists, and relevance to LDCs', *World Development*, **16**(3), 373–84.

Tisdell, C.A. (1990), *Natural Resources, Growth and Development*, New York, Westport, CT and London, UK: Praeger.

Tisdell, C.A. (2005), *Economics of Environmental Conservation*, 2nd edn, Cheltenham, UK and Northampton, MA, USA: Edward Elgar Publishing.

Tisdell, C.A. (2009), 'Poverty, policy reforms for resource-use and economic efficiency: neglected issues', *The Singapore Economic Review*, **54**(2), 155–66.

Tisdell, C.A. (2011), 'Core issues in the economics of biodiversity conservation', *Annals of the New York Academy of Sciences*, R. Costanza, Karin Limburger and Ida Kubiszewski (eds), **1219**(1), 99–112.

Tisdell, C.A. (2014), *Human Values and Biodiversity Conservation: The Survival of Wild Species*, Cheltenham, UK and Northampton, MA, USA: Edward Elgar Publishing.

Tisdell, C.A. (2015), *Sustaining Biodiversity and Ecosystem Functions: Economic Issues*, Cheltenham, UK and Northampton, MA, USA: Edward Elgar Publishing.

WCED (World Commission on Environment and Development) (1987), *Our Common Future*, Oxford: Oxford University Press.

Westlund, H. and F. Adam (2010), 'Social capital and economic performance: a meta-analysis of 65 studies', *European Planning Studies*, **18**(6), 893–919.

4. Values, economic valuation, and the assessment of environmental and economic change

4.1 INTRODUCTION

Assessing the economic value of environmental and economic change, especially its social economic value, is a challenging task. This is not only because empirical identification of the values and preferences held by individuals is not straightforward but because social valuation needs to take account of the fact that individuals often hold different and conflicting values and preferences. Social situations in which groups are willing to act as a team (Marschak, 1955), that is, have identical preferences, can occur but are unlikely to be common in a world dominated by the presence of self-interest.

A second problem is that the preference orderings and value systems of individuals about environmental and economic possibilities are often incomplete. Furthermore, they can display internal logical inconsistencies. Given bounded rationality, these possibilities must be very common. Nevertheless, it is usually irrational to fully specify preferences, and lack of complete specification and some logical inconsistencies in preference orderings do not rule out the possibility of rational (optimizing) behaviour (Tisdell, 1996, Chs 2 and 3).

A third complication is that dynamic (time-related) changes (of a varied nature) occur in individual preferences and in communal values thereby creating several difficulties for the valuation of environmental and economic changes. However, neoclassical economics pays little or no attention to the process of preference formation and change. Basically, it treats preferences as being exogenous to its modelling. Furthermore, it usually assumes that economic agents are well informed; their preference orderings are complete and that they do not display internal logical inconsistencies. While this theoretical approach can provide insights into the way markets operate, it is subject to some serious limitations as a basis for valuing major economic and environmental changes. Moreover, economic techniques (such as social cost–benefit analysis and economic impact analysis) only value economic and environmental changes from a limited ethical or normative perspective (as was recognized by Pigou, 1932) and the methods used for such valuations can have significant shortcomings. Nevertheless, they are often the best available methods and can be quite informative.

A fourth consideration is the extent to which the preferences of individuals ought to be the basis for the valuation of possibilities, that is, for deciding what is desirable.

This chapter elaborates on the above-mentioned issues. In turn, it considers different types of value and preference systems; problems involved in empirically determining the values and preferences of individuals; methods for social evaluation of economic and social change based on welfare economics; the relevance of measures commonly used in macroeconomics for measuring human well-being; and economic evaluation based on economic impact analysis rather than methods developed in welfare economics.

4.2   DIFFERENT TYPES OF VALUES AND PREFERENCE SYSTEMS

Economic approaches to decision-making and behaviour are dominated by the modelling of relationships between means and ends. Ends are usually specified by some type of preference ordering. These orderings are often an important element in both positivist and normative economic models. Positivist economic models are concerned with how economic agents act, whereas normative economic models mainly focus on how they should act given the preference of individuals for resource use. On the whole, economics pays scant attention to the worthiness of individual preferences per se. Nevertheless, in pursuing one of its objectives, economics cannot be value free, namely in its search for organizational ways to maximize (in a social setting) human well-being based on what individuals want subject to the limited resources available for this purpose.

There are several reasons why this is so. First, when personal preferences about resource use display conflict, interpersonal comparisons of benefits (utility) must be made to take account of this conflict. Secondly, while, on the whole, the democratic principle that the preferences of all should count underlies most welfare economics, individual preferences do not count equally in most economic systems, nor in economic valuation methods for resource use. Differences in the ability of individuals to pay make up one source of this inequality. When the willingness of individuals to pay is adopted as a measure of economic value, this ensures that the preferences of those on higher incomes receive greater weight than those on lower incomes. Thirdly, questions have been raised about whether the preferences of some individuals, such as the insane, should count at all. Several different factors (such as the nature of political systems and the geography of jurisdictions) limit the actual extent to which the preferences of individuals count in social decision-making. They determine whether they count at all as well as the way in which they count.

Economic models to determine what social choices should be made cannot be value free, although the values assumed are not always made explicit. This lack of transparency often occurs, for example, in the use of cost–benefit analysis.

The question of what values should be held by individuals has been discussed by philosophers. One approach is that of deontological ethics. It involves a search for moral imperatives or obligations that should apply to all human beings. These obligations are usually stated categorically.

Examples of this type of ethical approach include the following:

- It is wrong to kill sentient beings. Therefore, human beings should not kill them.
- All species have a right to exist and humans have an obligation to respect this right.
- Cruelty to animals is wrong. Therefore, humans are obliged not to be cruel to animals.

It is doubtful whether values of this type can be determined by rational means. Nevertheless, different approaches have been adopted to try to determine values which human beings ought to follow. Immanuel Kant (1964) tried to distil such values rationally. However, the only basic moral obligation he was able to distil was that all human beings should act with goodwill towards one another. He is thought to have believed that the intention of an act was more important ethically than its consequences (Anon, 2016a). Therefore, for example, action to save a *doomed* species or to prevent its loss locally may be morally justified. For instance, ineffective attempts to save the koala locally could be ethically justified given this deontological approach (cf. Tisdell et al., 2015).

Others rely for the determination of deontological values on such means as interpretations of God's will as expressed in holy texts, or depend on their own conscience. However, the latter is unlikely to be an independent entity. It is likely to be moulded by communal values and values transmitted in social contacts. Such contacts can be quite varied. They include family upbringing and the groups with whom individuals become associated.

A problem with categorical values is that they usually do not allow for exceptions and can result in fanaticism. This is of particular concern if these values are not well founded. They can result in actions and communal choices having negative social consequences.

Both the intrinsic worth of held values and their consequences are important. For example, the moral obligation not to steal and to pay debts promptly might be held to be justifiable from a social point of view because all will be eventually worse off economically if stealing and failing to pay debts become ingrained in society.

Mainstream economics is based on anthropocentric values. The only values or preferences which count are those that individuals hold. However, these values need not be entirely self-centred. They can include altruism (concern for the welfare of others) and can also incorporate other concerns such as those for the welfare of animals or for the existence of wild species. In economic cost–benefit analysis, monetary sums are derived in an attempt to quantify these preferences or values, for example based on willingness to pay or willingness to accept compensation criteria.

The relationship between communal values and economic behaviour and decisions has not been well explored by economists. Nonetheless, this relationship can be important. Moreover, it is significant that communal values have varied historically and may be dissimilar in different societies. In Western societies, avoidance of cruelty to animals has strengthened as a collective (communal) value with the passage of time and so has the purported obligation of humans to act as stewards of nature (Passmore, 1974). For example, baiting of bulls and bears was a pastime in Elizabethan England but today it is regarded as unjustifiable nearly everywhere, and the RSPCA (Royal Society for the Prevention of Cruelty to Animals) is very active in trying to stamp out cruelty to animals and their neglect.

The hunting of wild animals is socially less acceptable now than it used to be. In the 1920s, for example, koalas were hunted for their skins and some Australian states paid a bounty for each koala killed (Moyal and Organ, 2008). Deliberately killing koalas today is socially unacceptable and subject to legal penalties. Nevertheless, koala populations are declining in several parts of Australia primarily as a result of loss of their habitat, mainly due to urban expansion and increasing agricultural intensification (Tisdell et al., 2015).

Whether or not the desirability of social actions and choices should be determined by intrinsic values or by the desirability of their actual consequences, or by a combination of both is controversial. Let us consider some alternative views about this matter in relation to the social desirability of conserving natural biodiversity.

4.3   AN INTRINSIC APPROACH TO THE VALUE OF CONSERVING NATURAL BIODIVERSITY VERSUS CONSEQUENTIAL APPROACHES: SOME EXAMPLES

Aldo Leopold (1933; 1966) argues that human beings are just a part of the web of life and have no right to exterminate any of its parts, including species which are pests (for example, predators) from the point of view of human beings. Consequently, the utility to human beings of the continuing existence of species should not (in his opinion) be the basis for conserving or eliminating species. The primary goal of conservation ought to be to conserve the whole 'web of life' or, in other words, maintain the continued existence of natural ecosystems as an end in itself. Leopold's land ethic contrasts with approaches (such as those in the Millennium Ecosystem Assessment, 2003; 2005) that assume that decisions about biodiversity conservation should be based on their contribution to human well-being.

Leopold's approach also differs from those of Ciriacy-Wantrup (1968). Ciriacy-Wantrup also adopts an approach which is supportive of the conservation of all species. He argues that the cost of conserving each species at its minimum viable level of population can be expected to be low compared to the potential benefits to human beings of doing so. Therefore, it is wise to try to conserve all species. However, the cost of conserving all species in relation to their potential benefits to humans is not low. For example, the cost of conserving the orangutan is high (Tisdell, 2015, pp. 294–5). Furthermore, the minimum viable population concept has several important limitations (Hohl and Tisdell, 1993; Tisdell et al., 2015), and Ciriacy-Wantrup's approach is of little use if not all species can be saved due to resource constraints (Tisdell, 1990). In addition, it is impossible to conserve most species in isolation because the survival of many depends on the survival of other species and suitable ecological (environmental) conditions must be sustained in order to conserve each.

Kant's ethical position when applied to nature conservation indicates that it is ethically appropriate to try to save a species even when it is clear that its continuing existence is doomed. Acting with the intention of trying to conserve it is more important than the result. There is experimental evidence that some individuals are prepared to act in this way (DeKay and McClelland, 1996). The effort of some groups trying to conserve koalas in areas of Australia where their continued presence in the wild is doomed appears to be partly motivated by this ethic. This and other reasons for this type of behaviour are discussed in Tisdell et al. (2015).

4.4 CHALLENGES INVOLVED IN BASING COLLECTIVE PREFERENCES ON THE ECONOMIC VALUATIONS OF INDIVIDUALS

Economists pay a lot of attention to the empirical determination of the preferences of individuals for commodities, particularly unmarketed or partially marketed ones – these include a large set of public goods and quasi-public goods, such as environmental goods. This is motivated to some extent by the democratic principle that the preferences of every individual (with a few exceptions) ought to count in determining the collective supply of these commodities. Apart from the question of whether this principle should be rigidly followed, the empirical determination of individual preferences and economic valuation is subject to important limitations as are processes of deriving collective economic valuations from these individual valuations. It is important to consider the limitations involved because these procedures form the basis of social cost–benefit analysis and need to be kept in mind when assessing its reliability. This is not to suggest that such analysis is not worthwhile but, as Pigou (1932) pointed out, it only provides one perspective on the worth of economic and environmental changes.

4.5 AN INTRODUCTION TO METHODS USED TO DETERMINE THE ECONOMIC VALUES PLACED BY INDIVIDUALS ON THE AVAILABILITY OF UNMARKETED ENVIRONMENTAL COMMODITIES

Two basic approaches have been adopted by economists to determine the economic values which individuals place on the supply of unmarketed environmental commodities. These can be divided into:

- revealed preference methods; and
- stated preference approaches.

However, methods exist which make use of a combination of both these approaches.

In the case of revealed preference, the economic values or preferences of individuals are inferred from their economic behaviour. Examples include the use of travel cost methods and hedonic pricing. On the other hand, stated preference methods rely on questioning individuals about their economic valuations (for example, contingent valuation methods) or asking them to rank their preferences for alternative economic possibilities, for instance choice modelling methods.

Both revealed preference methods and stated preference approaches rely for valuation purposes on the assumption that the actual behaviours of individuals or their stated choices or values accord with the neoclassical economic model of human behaviour. Consequently, individuals are supposed to act to maximize their utility (preferred economic outcomes) subject to their availability of resources for achieving its aim. As typified by the concept of rational economic man, strong assumptions are made about the rationality of economic behaviour. While this concept provides some basic insights into the operation of markets, it faces greater limitations when applied to the economic valuation of unmarketed environmental commodities. In particular, the knowledge of individuals about the nature of unmarketed commodities is usually more deficient than in the case of marketed commodities (Tisdell, 2007).

In turn, revealed preference methods for determining the value of unmarketed environmental commodities will be assessed first and then the same will be done for stated preference methods. The following is not intended to be a complete detailed catalogue of the limitations of such methods. Its main purpose is to highlight limitations which have often received little attention in the literature. An excellent overview of several approaches to economic valuation (as well as other types of valuation) can be found in Pascual et al. (2010).

4.6 A GENERAL ASSESSMENT OF REVEALED PREFERENCE METHODS FOR DETERMINING THE ECONOMIC VALUE OF UNMARKETED ENVIRONMENTAL COMMODITIES

Travel cost methods (one type of revealed preference approach) have been widely used to measure the economic value of some recreational attractions. These can include man-made outdoor attractions and natural ones, or recreational sites containing both man-made and natural features. Travel cost methods use the cost of travelling to a recreational site (the entry to which may be free) as a proxy for the price of enjoying it for recreational purposes. Taking into account the fact that the distance individuals have to travel to visit a recreational site varies (and so, consequently, does their travel cost and therefore, the implicit price of their visits), a demand curve for visiting a site can be estimated based on these travel costs and the relative frequency of visits from the geographical areas from which visitors are drawn. In turn, this allows the economic surplus (consumers' surplus) to be calculated for visits to the focal site. This surplus is normally used as an indicator of how much a site is valued for recreational purposes. It is a measure of the economic value obtained by visitors over and above the cost of travelling to the site.

It should, however, be borne in mind that the economic value of visiting a site fails in many cases to measure its total economic value. It is normally just one component of that value. Consider a couple of illustrations.

Man-made dams are often built with multiple purposes in mind. These can include supplying water for use in agriculture and in urban areas, as well as regulating water flows for flood control. In addition, they and their surroundings are frequently used for recreational purposes such as boating, fishing, hiking, for picnics and as scenic attractions. The travel cost method was applied (at an early stage of its development) to determine the economic value of such dam sites (Knetsch, 1964). Politically, it helped to boost the economic case for the construction of dams. However, dams can have major adverse environmental and ecological consequences. These appear to have been given limited consideration in the early social cost–benefit analyses of the desirability of the construction of dams. Subsequently, travel cost methods were increasingly applied to the economics of conserving natural (or near natural) areas, such as national parks or natural conservation areas.

If the off-site economic benefits of natural conservation areas are high, then the economic value derived from visits (as determined by the travel cost approach, for example) will constitute a small portion of their total economic value. Off-site economic benefits can consist of existence, bequest and option values as well as hydrological benefits of various types (for example, less erratic run-off of rainwater, cleaner water in streams) and so on. There are, however, other aspects of the economic valuation of natural areas based on travel cost methods which can be overlooked. These include standard ones mentioned in the literature, such as problems arising when journeys are of a multipurpose nature or the actual journey itself is pleasurable or unpleasant. The pleasurability of the journey should be specifically estimated (see for example, in relation to tourist visits to Antarctica; Tisdell and Wilson, 2012, p. 145).

As is recognized in the literature but which is worth emphasizing, the recreational value of a site is often a function of the number of visitors to it. When this is so, the economic surplus which an individual gets from visiting a site depends upon the cost to the individual of visiting the site and the number of visitors. Ways in which the individuals' economic surplus from visiting an outdoor site can alter with its visitation rate include:

- Social effects – an individual may gain satisfaction purely from the presence or absence of others at a site. This satisfaction can vary in different ways. It may, however, be common for it to rise at first with the visitation rate and then decline.

- Crowding and congestion impacts.
- Damage to the environment and conservation value of the site caused by a rising number of visits.

As a rule, applications of the travel cost method do not take account of the effect on individual values of the total number of visits (visitors) to a site. Estimates are usually made given the existing aggregate number of visits. As a result, the potential economic value of visits to a site can be underestimated. This is illustrated in Figure 4.1. Curve OABC represents the total economic surplus individuals obtain from visits to a site as a function of the total number of visits, everything else held constant. Total economic surplus from visits is maximized when there are X_2 visits per period. Should the number of visits be greater than this, say X_3, the travel cost method will underestimate the potential economic value of the site for recreational purposes. Again, if there are fewer visits than X_2, say X_1, travel cost methods will not reveal the potential for the economic surplus to increase as the visitation rate rises.

Various policies can be adopted to respond to situations where the number of visits to a site reduces the total economic surplus obtained

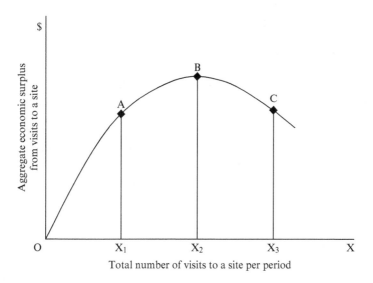

Figure 4.1 *An illustration of a valuation deficiency (explained in the text) when travel cost methods are used to estimate the aggregate economic surplus obtained from a site-specific attraction*

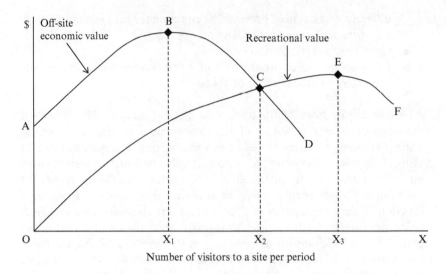

Figure 4.2   An illustration of the possible conflict between the economic value of visiting a site and its off-site (economic) conservation value

from visits. For instance, measures may be adopted to ration the number of visitors (for example, by the pricing of visits or by placing quotas on the number of visits), the infrastructure may be improved at the site designed in order to reduce environmental damage, and the areas used for different types of recreation may be separated (in order to reduce negative spillovers) when the site is used for multiple recreational purposes.

As indicated above, the economic value of visits to many attractions (such as outdoor environmental attractions and heritage sites) does not usually measure their total economic value. Furthermore, there can be conflict between maximizing the aggregate economic value obtained from visits to a site and the maximization of its off-site conservation value. This is illustrated by Figure 4.2 where the relationship ABCD represents the off-site conservation value obtained from the site and OCEF indicates the aggregate economic surplus derived from visits to the site. If the number of visits to the site exceeds X_1, the off-site economic benefits obtained from the site decline, although the economic benefits from visits continue to rise until the number of visits per period of time reaches X_3. The total economic value of the site is equal to its on-site plus the off-site benefits and reaches its maximum for a number of visits between X_1 and X_3. Maximizing the economic value of the visits to the site will not maximize the total economic value of the site. Furthermore, if one were to rely only on the value of visits to the site to estimate the site's total economic value, this will be an underestimate if the off-site value of the site is positive.

Note that the off-site economic valuation of a site may not be independent of visits to the site. Visiting a site can increase the willingness of visitors to pay for the conservation of the site if their experience is favourable. This can occur even though visits may cause some environmental damage to the site. Consequently, a conservation dilemma exists, namely, limiting visits to the site reduces environmental damage to the site but may lessen public support for the conservation of the site.

Visitors' lack of knowledge about the recreational value of a site can seriously restrict the applicability of travel cost methods. Application of travel cost methods is most appropriate to situations where individuals journey to a single site and have sound prior knowledge of the attractions of the site. This is likely to be the case if all or most visitors to a site visit it frequently. There are, however, many situations in which visitors to a site have little knowledge of its attractions before visiting it. This may be so if they have not previously visited it, or do not make much of an effort to gather information about it prior to their visit, or the information obtained is deficient in some way. Empirical evidence indicates that the knowledge of many individuals is poor when they visit significant outdoor recreational attractions for the first time (Tisdell and Wilson, 2012, Chs 7 and 8). This is a particular problem when visits can be classified as experiential goods (Tisdell, 2007). Many national parks and similar attractions are located a long way from home and visited by individuals once in their lifetime, or if more than once, only after a long interval of time during which these attractions may have altered or repeat visitors' memory of them may have become blurred. International tourism journeys in particular are likely to be subject to this problem. Furthermore, recreational journeys are often undertaken for multiple purposes, that is, not solely to visit a single site. Therefore, caution must be exercised in determining the type of situations to which the travel cost method can be applied even if data exists on travel costs and other relevant variables.

Hedonic pricing methods provide additional possible revealed preference-ways of measuring the economic value of environmental amenities. They rely on the willingness to pay to avoid unfavourable environmental conditions in order to enjoy more favourable ones. They do this by examining differences in the prices of commodities associated with variations in environmental quality. These variations occur usually (but not always) as a function of location. For example, residential land values often vary according to the access they provide to environments of different quality, for instance the extent to which they are exposed to varying degrees of environmental pollution (especially noise and air pollution), or the scenic views they command (Pearson et al., 2002). Differences in land prices can provide an indicator of how such variation in environmental quality is valued. However, prices of this type are usually influenced by a variety of factors and it can be difficult to disentangle the effects of individual influences, especially if multicollinearity occurs. For example, residing in an environmentally favourable area may be partly due to its environmental attractiveness but a 'snob' element can also be involved.

4.7   A GENERAL ASSESSMENT OF STATED PREFERENCE METHODS OF DETERMINING THE ECONOMIC VALUE OF UNMARKETED ENVIRONMENTAL COMMODITIES

There is a large set of environmental characteristics which cannot be valued using revealed preference methods. However, many of these can be valued by using stated preference methods, although the effectiveness of these methods and the implications of the values obtained are subject to considerable debate. These methods are designed to elicit, by questioning individuals, the maximum amount of money they would be willing to pay to retain the availability of a desired commodity (an environmental amenity, for example) or to be rid of an unwanted one, or to elicit the minimum amount they would have to be paid to forgo a wanted commodity (for instance, a desirable environmental feature) or to endure an undesired one.

Willingness to pay and willingness to accept compensation values can differ substantially (Knetsch and Sinden, 1984). It is a moral question of whether willingness to pay or to accept compensation should be the basis of individual resource allocation decisions based on a social cost–benefit analysis. Take, for example, the case of a factory emitting pollutants that adversely affect people in its neighbourhood. Is it more appropriate for social decisions about whether to regulate or control this pollution to be based on the willingness of the victims to pay to control this pollution or on their willingness to accept compensation to tolerate it? Should the polluter pay to pollute or be paid not to do so? Ethical considerations are required to answer these questions.

When willingness to pay is the basis of social decision-making about resource use, those with greater ability to pay (those on higher incomes) have a larger influence on the nature of environmental change than those with less ability to pay. This effect is less marked when willingness to accept compensation is the basis of social decision-making. However, even those on lower incomes are likely to be willing to accept a lower payment than those on higher incomes to tolerate a negative externality.

Contingent Valuation: Different Ways of Eliciting Values and Limitations of this Type of Valuation

Contingent valuation is one approach to eliciting the economic values which individuals place on hypothetical environmental or economic changes. There are two basic approaches to determining these values. These are:

- by asking an individual the maximum amount which he/she would be willing to pay to bring about a desired change or would need to be paid to tolerate an unfavourable change (this is a single-bid approach); or alternatively,
- engaging in a trial-and-error process in which the individual is presented with a sequence of monetary values with a view to determining these values by a step-wise process (multiple bidding).

The first procedure is less time-consuming than the latter but possibly the second procedure forces respondents to think more carefully about their economic valuation. It may, however, be subject to starting point bias, that is elicited values can vary according to the initial value trialled. It may also be influenced by the spacing between the tested values.

Other Potential Distortions in Economic Values Obtained by Applying Stated Preference Methods

Several additional distortions (see the non-exhaustive list in Table 4.1) can occur in the economic values elicited by contingent and other stated valuation methods. Little attention has been given to some of these distortions. For example, attention problems and drop-off effects are often neglected. In the former case, stated values can be distorted because the respondent is required to focus his/her attention on the value of the object to be valued. As a result, given the limited span of attention or concentration of individuals (an element of bounded rationality), the respondent is unlikely to consider fully opportunities forgone in proposing his/her willingness to pay for the conservation of the focal object.

Drop-off or erosion in stated values can occur with the passage of time following an individual's initial favourable (or unfavourable) experience or impression of the object to be valued, even if the lapse of time is the only significant change which occurs. Consequently, values elicited soon after a vivid experience with the object to be valued are likely to be different from those elicited at a later time (Tisdell, 2014, Ch. 2). For example, stated willingness to pay values are liable to be higher for favourable experiences

Table 4.1   Possible (general) distortions in economic values obtained by applying stated preference methods

Type or source of 'distortion'	Comment
Strategic bias	Values exaggerated to influence policy outcomes.
Answers reflect social values	As a result, stated values may not really accord with the respondent's actual values.
Answers intended to please interviewer	Most likely to be a problem in face-to-face interview.
Process regarded as 'academic' by respondents	Responses are not taken seriously especially if it is believed that compensation or payment is very unlikely.
Hypothetical bias	Respondents find it difficult to imagine the environmental changes or conservation proposals to be assessed.
Information limitations	Respondents may have little or no knowledge of the environmental possibilities to be assessed. If information is provided, it is usually selective and can therefore result in bias.
Time constraints	Respondents may lack time to carefully assess possibilities. Therefore, responses can involve inadequate consideration. The importance of this varies with the elicitation method used.
Attention bias	The process of elicitation results in the respondents' preoccupation with the issue raised and to the neglect of competing issues. This raises or depresses the elicited value. This has similarities with the Heisenberg effect in physics.
Drop-off effects	Other things being held constant, the initial valuation by an individual of an environmental commodity can alter with the passage of time as the individual's knowledge or experience with the commodity recedes in his/her memory. Favourable values may fall and unfavourable ones may rise.
Sympathetic (emotional) effects	For example, support for conserving wild species is biased in favour of those with humanoid features.
Historical changes in values due to alterations in communal values	Individual values are not independent of communal values. Communal values have varied with the passage of time.
Cultural differences in values	Cultural values differ between societies, resulting in different valuations by individuals and conflict between societies about these values
Whole-and-part bias	When individuals are asked to value part of an environmental phenomenon, their answers may reflect the valuation of the whole phenomenon.

when elicited not long after the experience than when obtained later, even if no material change in circumstances has occurred. It is therefore unclear whether the elicited values close to the experience or those elicited later should be adopted for policy purposes. Nevertheless, sometimes the same policy may be supported both by values elicited close to the favourable experience and by those obtained at a later time, as was found when considering the economics of conserving the mahogany glider (Tisdell, 2014, Ch. 6).

Considerable evidence exists that individuals value highly the continued existence of species with humanoid features (Tisdell, 2014, Chs 13 and 14). This sympathetic effect can result in insufficient value being placed on the existence of essential species for sustaining the web of life (complete ecosystems), the presence of which may also be necessary for the survival of favoured humanoid species.

As mentioned above, changes in communal conservation values can occur within a relatively short period of time (within just a few generations). These are usually reflected in alterations in individual valuations. For example, substantial changes have occurred in communal attitudes towards nature conservation in Yellowstone National Park since its foundation, as is evident from the May 2016 issue of the *National Geographic*. As pointed out in this issue, between 1881 and 1950 a high priority was placed on increasing fishing opportunities in the park. As a result, many non-resident species of fish were introduced and fish-eating predators, such as bears and pelicans, were killed. Significant ecosystem change occurred and native fish species were imperilled. However, now the priority is on protecting indigenous species. In general, there is much more emphasis today on the restoration and conservation of the 'original' natural ecosystem. This has included the controversial introduction of the grey wolf.

Significant cultural differences sometimes exist between societies and influence the stated values of their citizens. On the whole, for example, a higher proportion of Japanese appear to favour whaling than those living in countries such as the United States and Australia. At an earlier time, attitudes of both of the latter countries to whaling were entirely different – both were involved in whaling. Consider another example. Many groups of Australian Aborigines regard particular landforms as sacred sites and place a high value on their conservation, whereas some of these sites are little valued by other Australians. Similar differences in values exist between those of American Indians and the rest of the American population.

Problems are also likely to be encountered in ensuring that the sample (whose stated values are elicited) is representative of the relevant population being considered. Usually, available funds for determining stated values are quite limited. This restricts the sample size, there may be self-selection bias (only those with most interest in the object may respond), and it may be impossible to draw the sample in a random manner.

Allowing for Systematic Bias in Stated Values

If regular patterns of distortions occur in sets of stated values, these can be allowed for, to some extent, in investigations. If, for example, it is known that the individuals in the sample from whom stated conservation values are obtained have higher incomes and more education than the general population, then extrapolation of the results to the general population may need to be qualified or adjusted. Usually those on higher incomes and/or more education are willing to pay more for environmental conservation than those with less income and education.

Experimental evidence also indicates that, on the whole, individuals are willing to pay more on average for the conservation of well-known wildlife species than for those that are poorly known (Tisdell, 2014, Ch. 11). These averages tend to converge when knowledge of the less well-known species increases while at the same time greater dispersion occurs in values placed on the conservation of each of the species that were originally poorly known. It is therefore highly likely that the elicitation of stated values will undervalue (on average) the conservation of species that are poorly known or unknown.

Note that estimating the willingness of a population to pay for environmental conservation need not be the prime purpose of eliciting stated values. In some cases, the main purpose is to use experiments to determine how these values alter with changes in explanatory variables, such as alterations in the degree and nature of information available to respondents.

The Reliability of Stated Values for Experiential Environmental Commodities

Some environmental commodities are of an experiential nature and may only be experienced once in a lifetime or again only after the lapse of a long period of time. The nature of their value, therefore, has much in common with that of movies, novels and so on. This seems to be mainly true of visits to see natural and man-made wonders far from home. The main purpose of such visits may be to satisfy curiosity. However, even if more frequent visits to an environmental site are on the individual's agenda, the individual's first visit is of an experiential nature. Further visits may depend on the experience and knowledge obtained. Subsequent visits may also depend on accumulating experience and knowledge. It is, however, one-off or initial visits that call for particular caution in drawing conclusions from stated values obtained from visitors about an attraction.

In this regard, one should distinguish between the stated value of the visit and the stated value of conserving the attraction. Consider the maximum willingness to pay for a visit. This could be elicited prior to the individual's visit or following it. The prospective and the retrospective values are likely to differ. The retrospective value could be sought by asking visitors to state the maximum amount they would have paid to visit (for the first time) *given* their experience of the environmental facility. A hypothetical problem is posed and the answer will depend on the length of time between the experience and the elicitation. Dissimilar prospective and retrospective stated values result in differing estimates of consumers' surplus. Therefore, a decision has to be made about which of these values is most relevant for policy purposes.

One could also enquire about willingness of visitors to pay for subsequent visits. This may be low or zero after first-time visits, especially if those who come to the site basically do so to satisfy their curiosity. On the other hand, some visitors may be considering using the site regularly for recreational activities and if it is found to be favourable for this purpose, their willingness to pay for future visits may be high.

The above refers to the value of visits, that is, to use value. Visits, however, may also influence non-use values, and these may be higher or lower than prior to a visit. An individual's stated willingness to pay to keep an environmental attraction in existence should reflect the sum of his or her assessment of its use and non-use values. Prospective and retrospective total values may differ considerably, especially before and after the first visit to a site. For example, for some sites many individuals may not want to visit the site again for recreation (future use value is low or zero) but their first visit may substantially elevate the value they place on the non-use value of the attraction. The willingness of an individual to pay for future visits may be low or zero even if the visitor's first visit to an environmental attraction has given the visitor great satisfaction. All of the above factors complicate the empirical valuation of experiential commodities, several of which are environmental attractions.

4.8 CRITERIA FOR COLLECTIVE CHOICE BASED ON INDIVIDUAL PREFERENCES: SOME WELFARE ECONOMICS APPROACHES

Several different approaches to determining collective preferences exist in welfare economics. The least controversial one is the Paretian improvement criterion. An economic or environmental change is considered to be desirable if it makes some individuals better off without making any worse off. It has been criticized on the basis that it favours the status quo based on the distribution of income or wealth and that few changes may be possible which satisfy this criterion. The potential Paretian improvement criterion (also known as the Kaldor–Hicks criterion) is much more often applied, particularly in social cost–benefit analysis.

It assumes that an economic or environmental change is desirable if the gainers from it could compensate the losers and be no worse off than before the change, even if actual compensation is not paid to the losers. The question of whether and how much compensation should be paid to the losers is an ethical one. However, the transaction costs involved in making such a payment also need to be kept in mind. If these are high, then gainers may not really be able to compensate losers after these are taken into account. If justice dictates that compensation should actually be paid then, in cases like this, because gainers are unable to compensate losers after the deduction of transaction costs, the change would not be socially desirable.

If the Kaldor–Hicks criterion is satisfied by a proposed change but compensation is not going to be paid to losers, one might want to consider whether any negative consequences for the distribution of income of the change outweigh the aggregate increase in net benefits otherwise obtained. This could be done by the social decision-maker by weighting the monetary gains and losses by the decision-makers' degree of concern for their impact on the welfare individuals affected. If the sum of these weighted values is positive, the change may be considered to be socially desirable and it could be judged to be undesirable if that sum is negative. The situation where all the losses and gains are given a weight of one (the Kaldor–Hicks case) is a special case of this type of welfare function. These types of weighted functions can be regarded as examples of a Bergson welfare function.

The Bergson Welfare Function

Bergson (1938) allows for different ways of determining collective preference functions for allocating resource use. It is probably inappropriate to call these collective or aggregate preference functions welfare functions, because they may, for example, reflect the preferences of a dictator, an oligarchy or of some other group in society. These orderings are assumed to be transitive and complete, and can be used in principle to determine the optimal allocation of resources, that is, optimal given the proposed ordering.

A hypothetical example is illustrated in Figure 4.3. This enables the

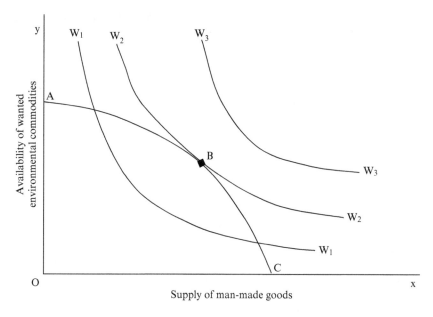

Figure 4.3   An illustration of Bergson's analysis of optimal collective choices

optimal trade-off between the availability of wanted environmental commodities and the supply of man-made goods to be considered. The
trade-off frontier is indicated by the transformation curve ABC and
selected collective indifference curves are identified as W_1W_1, W_2W_2 and
W_3W_3. Given this representation, the combination corresponding to
point B is optimal. If increased supplies of wanted environmental and
man-made goods are desired, it is optimal to be on the trade-off frontier.
Optimal positioning on this frontier will depend on the nature of the
'collective' (Bergson) preference function.

Bergson's analysis raises a couple of important issues. First, political
decisions about resource use may not rely on a mapping of individual
preferences to obtain social welfare functions. Secondly, it may not
be possible to achieve such a mapping that fully respects individual
preferences, as is demonstrated by Arrow's *Impossibility Theorem* (Arrow,
1951). There is also a third problem: the types of functions indicated in
Figure 4.3 may be poorly defined in practice or not even defined at all.
In reality, social choices are determined by political interactions and are
subject to institutional constraints. Estimates obtained by using economic analysis are just one possible influence on the outcomes of these
interactions.

4.9   MACROECONOMIC INDICATORS OF CHANGES IN HUMAN WELL-BEING

There are now many different macroeconomic and related indices of human well-being. This is, for example, apparent from the Wikipedia entry for the Human Development Index (HDI) (Anon, 2016b), which, apart from providing information about this index, provides a link to other macro-indicators of human well-being. It is not possible to review all of those here. Only brief notes will be provided about some of these indicators.

Note that none of these indicators are derived explicitly from individual preferences. The variables in these indices and their weighting reflect dominant community concerns about socio-economic change. For the most part, they might be regarded as portraying Bergson-type welfare functions. The available set of these indices has evolved with the passage of time. Changing communal values, greater availability of data and the altering nature of macroeconomics have influenced the type of indices developed. They are frequently used to make international comparisons of welfare and changes in welfare with the passage of time. Several of these are estimated and published by international bodies such as the United Nations Development Programme.

While these indices can be useful, a real danger is that their limitations are not always appreciated. It is easy to be deluded by the apparent quantitative precision of the indicators briefly considered below, namely gross domestic output per capita (GDP), gross national income per capita (GNI) and the HDI.

GDP and GNI per capita

Beginning in 1944, first gross national product (GNP) and subsequently, beginning in the 1980s, GDP were adopted as prominent measures of the level of aggregate economic activity of nations. GDP came to be preferred because it measured the aggregate level of total production within a country's borders. Aggregate employment in an economy is normally expected to be positively related to the level of its GDP, at least in the short run. Therefore, from a Keynesian perspective, it seems more relevant than GNP when assessing the aggregate level of economic activity. Nevertheless, GDP per capita can be a very poor indicator of the level of economic welfare, even if sustainability and environmental considerations are ignored.

A simple expression for the composition of GDP (Y) is:

$$Y = C + I + G + (X - M) \tag{4.1}$$

where C represents the level of final consumption, I is the level of private investment, G represents the level of government expenditure on goods and services, X is the level of exports and M represents the level of imports. In some countries, the economic benefits of government expenditure can be quite low because of corruption and inefficiencies in the distribution of this expenditure and there may be 'excessive' military expenditure. Trying to maintain a high trade surplus may also result in domestic consumption being lower than it need be. In the past, some communist countries engaged in forced savings and investment and this reduced economic welfare (at least in the short run) and there were serious shortcomings in the economic efficiency of their production. Consequently, the value of their estimates of the level of aggregate economic activity was even more problematic as an indicator of welfare than in market economies.

GNI is a measure of the level of aggregate income available to residents of a nation. It consists of income earned at home plus that obtained from abroad less any payments due to overseas residents or entities. It is equal to GDP plus income receipts from the rest of the world less payments to the rest of the world. GNI per capita is usually regarded as a better indicator of the level of welfare of a nation than GDP per capita and is an important variable included in the calculation of the human development indices of nations. Nevertheless, it remains a poor indicator of welfare. For example, if estimated by adjusting GDP for income payments to and receipts from the rest of the world, it is subject to the deficiencies associated with GDP values. Among other things, neither GDP nor GNI take account of the value of unmarketed environmental goods, the welfare costs of environmental pollution and degradation, the sustainability of income or consumption per capita, and the distribution of income. Furthermore, these values are also influenced by accounting *conventions*. A fairly informative outline of the limitations of these concepts and their measurement can be found in Anon (2016c).

The Human Development Index (HDI)

The HDI is computed by taking account of income per capita (GNI per capita), an education variable (mean years of schooling and expected years of schooling) and average life expectancy at birth. Each of these components is estimated as an index and the HDI is the geometric mean of these values. It can range between zero and one. A somewhat useful detailed account of HDI is available in Anon (2016b). The HDI concept was developed at the request of the United Nations Development Programme with the intention of providing a better indicator of human well-being than GDP per capita or GNI per capita. Despite the limitations of HDI values, higher HDI values are widely regarded as desirable and are seen as heralding an increase in human well-being. Given the precise quantitative estimates available for HDI, there is a risk that these values will be used indiscriminately in policy-making and econometric modelling to indicate changes in human well-being.

Limitations of HDI include:

- Life expectancy does not take account of the quality of life.
- Increasing schooling is not the only way of developing the capability of individuals, that is, allowing individuals to achieve their full potential. For some, it is not even optimal. Only years of formal schooling are considered and not the quality and the economic value of schooling.
- The standard HDI calculation does not take account of the distribution of income.
- It implies questionable 'welfare' trade-offs. For example, it implies that a reduction in the length of life can be compensated by an increase in GNI per capita, thereby leaving the value of HDI unchanged. In other words, the implicit social welfare function (sets of indifference curves) obtained using the HDI can be problematic.
- No allowance is made for the distress likely to be caused by involuntary unemployment and underemployment, and for changes in the economic security of individuals. In the latter respect, job security and the existence of social safety nets are important.
- The extent to which equality of opportunity exists is poorly proxied. The education variable is, however, intended to portray this.
- The extent to which poverty exists is not incorporated in the index.

The main problem is that it is unrealistic to expect a single macroeconomic variable (index) to measure adequately human well-being and differences in well-being between nations and changes in welfare with the passage of time. Several indicators must be considered in order to better access such differences and changes.

Allowing for Sustainability and Environmental Issues

The use of GDP per capita, GNI per capita and HDI as proxies for the level of human well-being has also been criticized for its failure to take account of the sustainability of these variables. High levels of current GDP, GNI, HDI and well-being may be obtained at the expense of their future levels by running down the stock of natural capital or transforming it in unfavourable ways. Furthermore, impacts on pollution of economic activity (such as the occurrence of air, water, noise and soil pollution) are not adequately allowed for.

It has become increasingly accepted that the level and sustainability of aggregate economic production and human well-being depend (among other things) on the level and composition of capital where capital is more broadly defined than in the past. In the past, capital was commonly defined as the *produced* means of future production. There was almost exclusive emphasis on measuring the level of manufactured or man-made capital and its contribution to economic development. It is now increasingly recognized that a broader concept of capital is needed in order to assess the future potential for economic production and welfare. There is now emphasis on the concepts of comprehensive (World Bank, 2011) or inclusive (Arrow et al., 2012) capital. The major components of comprehensive or inclusive capital are:

- manufactured or man-made capital;
- human capital;
- social capital; and
- natural capital.

It is extremely difficult to assign accurate economic values to these components because the stock of each of these types of capital is difficult to define and measure. The problem is magnified because their consequences for production and human well-being are not additive. Nevertheless, clearly an 'imbalance' between the components can result in human well-being being less than it could be and it being less sustainable than is desired. Because of variations in methods used to measure the components of comprehensive or inclusive capital, opinions and estimates can diverge considerably about how sustainable the income levels or consumption levels of nations are, given their existing stocks of these forms of capital (Engelbrecht, 2016).

The sustainability of income per capita and consumption per capita depends not only on the level and nature of initial stocks of the various types of capital but also on how well these stocks are managed during the period in which it is planned to sustain income and consumption levels. A further complication is that as a result of international trade, the sustainability of the levels of income and consumption of many nations depends on their access to the capital of other nations, for example, the stock of natural resources of other nations. Therefore, it can be misleading to consider only the adequacy of the stock of comprehensive or inclusive capital within a nation's borders when assessing the sustainability level of per capita income or per capita consumption. For instance, China's dependence on access to the stock of natural resources has become increasingly important to it as a means of sustaining Chinese income levels. Considerable disagreement also exists among economists about the extent to which forms of capital can be substituted one for another in order to ensure that future income levels are not less than currently. This disagreement is especially marked in relation to the extent to which the substitution of manufactured capital for natural capital is compatible with the goal of achieving sustainable economic development (see, for example, Tisdell, 1997).

4.10 LOCAL AND REGIONAL ECONOMIC IMPACT ANALYSIS

A popular means of valuing economic changes in local or regional areas is by relying on economic impact analysis. The purpose of such analysis is often to measure the impact on the level of income and employment in a region of a particular economic event, compared to its absence. However, as suggested by Anon (2016d), several other indicators may be used.

The usual starting point of this type of analysis is the expenditure associated with a new project or event, for example, a new mining project. A portion of this expenditure can be expected to flow to the local economy with the remainder (the leakage) being directed elsewhere. The greater the leakage, the smaller the economic impact on the local economy. Expenditure arising directly from the project will also have not only its initial impact but subsequent economic impacts (for example, impacts on local employment) depending upon how much of the income received initially from the project is spent on locally supplied goods and services. These subsequent economic impacts (multiplier effects) will be smaller, the greater is the economic leakage involved in supplying commodities locally.

Traditional economic impact analysis only takes account of the transactions for which monetary exchange occurs. It therefore ignores values associated with unmarketed goods. Furthermore, such evaluations often fail to take adequate account of opportunities forgone. For example, if the aim of policy is to provide the maximum boost to economic activity (for instance, employment) in a local economy, the employment or income-enhancing impacts of alternative possible projects may not be assessed. Furthermore, the economic benefit of helping residents of a local community where productivity and employment is low to move to an area where they can gain employment or higher levels of income may not be considered. In addition, political representation by regions can favour government support for employment and income-generating projects in marginal electorates even when these are not in the general economic interest of a nation.

Additional matters that need to be considered when assessing regional development projects include the following:

- For how long will the project inject expenditure into the local economy?
- Is the stream of expenditure injections likely to be stable or unstable?
- To what extent will the project employ locals compared to non-locals?
- What effect will the project have on the social stability of the local community?
- How acceptable are the environmental consequences of the development?

These can all be important issues. For instance, they are all important questions in relation to the regional mining developments in Australia. Although economic impact analysis can be undertaken purely as a positive exercise, it is frequently used to judge the desirability of economic policies. The basis of economic impact analysis and its indication of what types of economic policies might be desirable relies on quite different underpinnings to those of traditional welfare economics. Consequently, their policy implications are frequently in conflict, as is illustrated in Tisdell (2012).

4.11   CONCLUDING COMMENTS

The sample of economic valuation approaches and measures considered in this chapter indicate that they are quite varied. Each only provides a restricted measure of factors influencing human well-being. It is also important not to be deceived by the apparent precision of many of these measurements. Furthermore, all incorporate underlying normative assumptions and the acceptability of these needs to be considered if they are to be used for policy purposes. Moreover, it should also be borne in mind that special interest groups, including politicians, can be expected to make use of the most favourable economic valuation measures for supporting their particular aim. Therefore, one has to be careful in assessing the acceptability of claims based on the selection made.

REFERENCES

Anon (2016a), 'Deontological ethics', *Wikipedia*, accessed 17 March 2016 at https://en.wikipedia.org/wiki/Deontological_ethics.

Anon (2016b), 'Human Development Index', *Wikipedia*, accessed 18 May 2016 at https://en.wikipedia.org/wiki/Human_Development_Index.

Anon (2016c), 'Gross domestic product', *Wikipedia*, accessed 19 May 2016 at https://en.wikipedia.org/wiki/Gross_domestic_product.

Anon (2016d), 'Economic impact analysis', *Wikipedia*, accessed 23 May 2016 at https://en.wikipedia.org/wiki/Economic_impact_analysis.

Arrow, K.J. (1951), *Social Choice and Individual Values*, New York: John Wiley.

Arrow, K.J., P. Dasgupta, L.H. Goulder, K.J. Mumford and K. Oleson (2012), 'Sustainability and the measurement of wealth', *Environment and Development Economics*, **17**(03), 317–53.

Bergson, A. (1938), 'A reformulation of certain aspects of welfare economics', *Quarterly Journal of Economics*, **52**, 310–34.

Ciriacy-Wantrup, S.V. (1968), *Resource Conservation: Economics and Policies*, 3rd edn, Berkeley, CA: Division of Agricultural Science, University of California.

DeKay, M.L. and G.H. McClelland (1996), 'Probability and utility components of endangered species preservation programs', *Journal of Experimental Psychology: Applied*, **2**(1), 60–83.

Engelbrecht, H-J. (2016), 'Comprehensive versus inclusive wealth accounting and the assessment of sustainable development: an empirical comparison', *Ecological Economics*, **129**, 12–20.

Hohl, A. and C.A. Tisdell (1993), 'How useful are environmental safety standards in economics? The example of safe minimum standards for protection of species', *Biodiversity and Conservation*, **2**(2), 168–81, reprinted in C.A. Tisdell (2002), *The Economics of Conserving Wildlife and Natural Areas*, Cheltenham, UK and Northampton, MA, USA: Edward Elgar Publishing.

Kant, I. (1964 [1785]), *Groundwork of the Metaphysic of Morals*, H.J. Paton [trans.], New York: Harper and Row.

Knetsch, J.L. (1964), 'Economics of including recreation as a purpose of eastern water projects', *Journal of Farm Economics*, **46**(5), 1148–57.

Knetsch, J.L. and J.A. Sinden (1984), 'Willingness to pay and compensation demanded: experimental evidence of an unexpected disparity in measures of value', *Quarterly Journal of Economics*, **99**(3), 507–21.

Leopold, A. (1933), *Game Management*, New York: Scribner.

Leopold, A. (1966), *A Sand County Almanac: With Other Essays on Conservation from Round River*, New York: Oxford University Press.

Marschak, J. (1955), 'Elements for a theory of teams', *Management Science*, **1**(2), 127–37.

Millennium Ecosystem Assessment (2003), *Ecosystems and Human Well-being: A Framework for Assessment*, Washington, DC: Island Press.

Millennium Ecosystem Assessment (2005), *Ecosystems and Human Well-being: A Synthesis*, Washington, DC: World Resources Institute.

Moyal, A. and M. Organ (2008), *Koala: A Historical Biography*, Collingwood, Victoria: CSIRO Publishers.

Pascual, U., R. Muradian, L. Brander, E. Gómez-Baggethun, B. Martín-López, M. Verma et al. (2010), 'The economics of valuing ecosystems services and biodiversity', in P. Kumar (ed.), *The Economics of Ecosystems and Biodiversity: Ecological and Economic Foundations*, London, UK and Washington, DC, USA: Earthscan, pp. 184–256.

Passmore, J.A. (1974), *Man's Responsibility for Nature: Ecological Problems and Western Traditions*, London: Duckworth.

Pearson, L.J., C. Tisdell and A.T. Lisle (2002), 'The impact of Noosa National Park on surrounding property values: an application of the hedonic price method', *Economic Analysis and Policy*, **32**(2), 155–71.
Pigou, A. (1932), *The Economics of Welfare*, 4th edn, London: Macmillan.
Tisdell, C.A. (1990), 'Economics and the debate about preservation of species, crop varieties and genetic diversity', *Ecological Economics*, **2**(1), 77–90, reprinted in C.A. Tisdell (2002), *The Economics of Conserving Wildlife and Natural Areas*, Cheltenham, UK and Northampton, MA, USA, Edward Elgar Publishing.
Tisdell, C.A. (1996), *Bounded Rationality and Economic Evolution*, Cheltenham, UK and Brookfield, VT, USA: Edward Elgar Publishing.
Tisdell, C.A. (1997), 'Capital/natural resource substitution: the debate of Georgescu-Roegen (through Daly) with Solow/Stiglitz', *Ecological Economics*, **22**(3), 289–91, reprinted in C.A. Tisdell (2003), *Ecological and Environmental Issues*, Cheltenham, UK and Northampton, MA, USA: Edward Elgar Publishing.
Tisdell, C.A. (2007), 'Knowledge and the valuation of public goods and experential commodities: information provision and acquisition', *Global Business and Economics Review*, **9**(2/3), 170–82.
Tisdell, C.A. (2012), 'Economic benefits, conservation and wildlife tourism', *Acta Turistica*, **24**, 127–48.
Tisdell, C.A. (2014), *Human Values and Biodiversity Conservation: The Survival of Wild Species*, Cheltenham, UK and Northampton, MA, USA: Edward Elgar Publishing.
Tisdell, C.A. (2015), *Sustaining Biodiversity and Ecosystem Functions: Economic Issues*, Cheltenham, UK and Northampton, MA, USA: Edward Elgar Publishing.
Tisdell, C.A. and C. Wilson (2012), *Nature-based Tourism and Conservation: New Economic Insights and Case Studies*, Cheltenham, UK and Northampton, MA, USA: Edward Elgar Publishing.
Tisdell, C.A., H. Preece, S. Abdullah and H. Beyer (2015), 'Parochial conservation practices and the decline of the koala – a draft', *Economics, Ecology and the Environment*, Working Paper 200, Brisbane: School of Economics, The University of Queensland.
World Bank (2011), *The Changing Wealth of Nations: Measuring Sustainable Development in the New Millennium*, Washington, DC: The World Bank.

5. Social embedding: its nature and role in determining our economic and environmental future

5.1 INTRODUCTION

Social embedding limits the ability of humans to influence their economic and environmental future. Its presence locks societies into development paths which they can only alter to a limited degree. Scholars differ in their views about how strong this form of path dependence is. For example, Gowdy and co-authors (Gowdy and Krall, 2013; 2014; 2015; van den Bergh and Gowdy, 2009; Wilson and Gowdy, 2015) consider it to be very strong in modern societies, and Polanyi (1944) stresses its importance in 'primitive' societies. On the other hand, many European economic thinkers in the nineteenth century were optimistic about the ability of their societies to adopt socially desirable changes as a result of rational discourse. Although increasing scepticism has been expressed in modern times about the relevance of Utopian types of economics and the ability of societies to implement socially desirable policies (that is, ones considered to be desirable based on evidence and rationalism), very strong assumptions are still being made about the influence of economic rationalism on socio-economic changes, for example, as an explanation of why some societies adopted agriculture and others did not do so in prehistoric times. For instance, Weisdorf (2005) relies on neoclassical economic rationalism to explain why social groups decided to adopt agriculture or to continue to be hunters and gatherers.

Tisdell and Svizzero (2017) are critical of Weisdorf's approach given the varied and myopic nature of much human decision-making and the occurrence of different types and degrees of social embedding. On the other hand, the view of Gowdy and others that human societies exhibit the same type of ultrasociality as some species of communal insects (such as particular species of ants and termites) seems to be an exaggeration.

The causes and extent of social embedding can be very diverse. Therefore, the first objective of this chapter is to identify sources of social embedding and their characteristics. Both structural and cultural sources of embedding are considered. This matter is discussed and followed by the identification of forms of social embedding and their consequences in modern economies. The proposition is then considered that the evolution or development of economic systems alters social relationships and the nature of social embedding. Subsequently, two further matters are examined, namely whether social embedding is more marked now than in the past, and whether the analogy between the social embedding of human beings and that of communal insects is realistic.

5.2 STRUCTURAL EMBEDDING

Although structural and cultural social embedding are not always independent, it is helpful analytically to initially examine these as being separate features of social embedding. Economists in particular have given much attention to the way in which the structure of social decision-making influences collective outcomes when individuals pursue their own self-interest. Institutional structures often result in social embedding. For example, market systems create social embedding of a structural nature. The social desirability of this form of embedding and the economic structures required to get the most favourable collective outcomes from competition between individuals and entities immersed in market systems have been the subject of much debate. Support for such systems has mostly relied on the idea that they are very effective in promoting economic efficiency and stimulating economic growth, and consequently they enable the wants of humans to be more fully satisfied than alternative economic systems. Some supporters of market systems also consider these (or more competitive forms of these) to be desirable because they are supportive of personal liberty. Nevertheless, it is clear that market systems are not perfect from any of these points of view for reasons well covered in the available literature, for example, because of the presence of effects due to market failures of various kinds, such as environmental spillovers.

A group of structural social decision-making problems exist which have been characterized by the prisoner's dilemma problem. As is well known, given the structure of this problem, the pursuit of individual self-interest results in the worst possible outcome for all. Many situations exist in which selfish behaviour results in a socially inferior outcome for all those affected by such behaviour. The occurrence of such anti-social behaviour can be reduced by penalizing those engaging in it, including subjecting them to social disapproval. However, there are situations in which the social structure is such that penalties cannot be imposed on those engaging in anti-social behaviour or this behaviour cannot be effectively monitored. For example, so far the international community has not been able to agree on concrete methods to penalize nations which do not adequately reduce their GHG emissions. Furthermore, because nations self-report their progress in reducing these emissions, their reports may not always be accurate – each has an incentive to overstate its progress in this regard.

Obtaining social and collective economic changes wanted by human beings involves transaction costs because it is necessary to convince relevant members of the community of the desirability of such changes and to induce them to take steps to bring these changes about. If the costs of initiating desirable social change must be met by its initial advocates, their costs may far exceed their personal benefit from the change, and consequently it may never occur. The governance rules and customs developed by a society influence the distribution of costs which are borne by members of the community seeking social changes and therefore influence the possibility of such changes occurring. Consequently, a high degree of path dependence exists due to a combination of structural and cultural factors.

Olson (1965) has identified a range of structural factors that result in social embedding when individuals or entities pursue their own self-interest. Some have also been considered by Tisdell (1996, Ch. 8). In addition, socio-economic changes which must be executed on a large scale rather than on a small scale or incrementally in order to be successful can be an important barrier to such developments (Tisdell and Svizzero, 2017). Furthermore, the momentum for socio-economic change can be limited by myopia (Pigou, 1938) and uncertainty about the future. This can limit proactive behaviour in socially preparing for adverse changes, such as climate change (Weber and Stern, 2011). Ross et al. (2016) identify a variety of structural and cultural factors in contemporary communities which have created social barriers to responding to climate change.

5.3   CULTURAL EMBEDDING

The classification of different ways of organizing societies suggested by Weber (1947) is well worth considering because it can be related to cultural embedding and has been adopted by some economists. The suggested ways are:

- by command;
- by custom or code; and
- by market methods.

Most societies have relied on a combination of more than one of these approaches for their economic and social organization. Furthermore, their relative importance in organizing societies has varied historically and has differed between contemporaneous societies. For example, with economic development and growth, market systems have increased in importance as a means of social and economic administration. However, the operations of market systems also depend on elements of command and custom, and are subject to legal systems.

Economic entities comprised of several persons are the major producers in modern market economies, and authoritarian hierarchies usually determine the type of economic choices made by these entities and the role of employees in their organization. Laws made by governments are a type of command. Social norms or customs are also relied on for the efficient operation of market systems. These include respect for property rights, willingness to honour economic contracts and the payment of debts in a timely fashion. Laws provide a back-up to these norms. The less a society has to rely on legal systems to enforce these social norms the greater is likely to be the economic efficiency of the market system. Therefore, it can be advantageous to instil these values into members of a society by education and other means.

Two main forms of cultural embedding can be identified. These are the instilling of *social values* or norms in members of the community, and the propagation of *shared beliefs*. These may be used to reinforce one another. Religious texts, for example, usually propagate a set of beliefs, and within this context they advocate particular social values. This approach has been used (among others) to legitimize the rule of dominant groups or persons, for example the divine right of kings to rule and the mandate of heaven legitimizing the rule of Chinese emperors. Today, dominant groups use the media to propagate the values and beliefs favoured by them. In many countries, these communications extol the value of economic growth, the worth of market systems and the values which underpin these systems.

Types of social embedding of economic significance can be quite diverse. Consider some examples. Take for example, the filling of social and occupational roles in societies. These may be determined by any of the following or combination of these:

- by birth, that is by the social status of one's parents, as in medieval Europe or in the caste system in India;
- by market-related competition;
- by forceful or violent contest;
- by command or by the choice of those with social or political power.

This list is not exhaustive.

The distribution of commodities to individuals can also be determined in a variety of ways. These include the following:

- market mechanisms;
- command of dominant groups in society, as was largely the case in the USSR, and in Communist China prior to its market reforms;
- social rules: these were used by many hunting and gathering tribes (for example, Australian Aborigines) to share their food supplies;
- religious obligations.

Consumption patterns can also be subject to persistent and varied social influences. These include:

- Prohibitions (by governing authorities) on consumption of some commodities such as some types of drugs and other substances; for example, measures restricting the smoking of tobacco and, in some countries, the drinking of alcohol.
- Religious-based rules; for example, the avoidance of eating pork by followers of Islam and Judaism, and similarly, the avoidance of beef by most Hindus. The followers of some religions will not consume any animal-based products that require slaughtering an animal.
- In the past, the type of clothing individuals were allowed to wear was determined in several societies by their social status and their role in these societies.

All of these factors contribute to social lock-in, and consequently magnify economic embedding. The questions of why these types of embedding evolve and what determines their persistence are interesting ones. These questions cannot be fully answered here. However, we know that in some cases, social values can change relatively quickly, whereas in other situations they change very slowly or only as a result of catastrophic events or major socio-economic shocks. Nevertheless, the nature of economic systems is a strong force influencing types of social lock-in and their persistence. Let us consider types of social embedding in modern societies and subsequently their influence on economic development and social structures.

5.4   TYPES OF SOCIAL EMBEDDING IN MODERN (MARKET) ECONOMIES

The more costly or difficult it is for individuals or groups of individuals to alter an existing socio-economic system, the more likely they are to support it and to try to maximize their benefit from it. Given the current pervasive nature of the market system and the institutions supporting it, altering it radically would not be easy. Consequently, despite its shortcomings, there is strong political support for it and for economic policies designed to 'improve' its operations, for example, to promote economic growth in order to sustain or increase employment.

The Economic Growth Treadmill

Although widespread concern exists about the state of the environment, politically avoiding significant involuntary unemployment remains a high communal priority. The most common way of trying to achieve full employment is to adopt means to maintain or increase economic growth sufficiently to ensure a low level of involuntary unemployment. The greater is the rate of population growth and the more marked is the reduction in the employment of labour, for example, due to technological change or increased economic efficiency, the higher is the rate of economic growth required to sustain full employment.

Most economic entities see considerable advantages in continuing economic growth. For the employed, economic growth adds to their job security and is likely to increase their real income. For the involuntarily unemployed, it increases their chance of finding a job (or more hours of work) and obtaining an adequate income. Investors also favour economic growth because it is associated with enhanced aggregate demand and is seen as having positive consequences for the return on capital (profitability). It is, therefore, not surprising that trade unions and organizations of employees place considerable store on economic growth as an objective. It is difficult for governments to pursue environmental policies unless these can be shown to have positive employment and economic growth consequences. Consequently, modern market economies seem to be locked into a pro-growth mentality because of their macroeconomic structure (Tisdell, 1999, Ch. 6).

The idea also that the pursuit of individual self-interest is likely to result in the greatest collective good is perpetuated by the existence of the market system. Because the market system allows considerable scope for the exercise of self-interest, those wanting to reform the system usually have to bear the burden of proving that intervention in the system is justified. This adds to the difficulties of altering this system.

Committees, Remoteness in Governance, Knowledge and Social Responsibility

Whether or not the use of committees for making socio-economic decisions is more marked now than in the past is unclear. However, this form of governance tends to reduce individual responsibility for decisions, some of which may have adverse social consequences. The committee rather than the individual is liable to be blamed for unpopular committee-based decisions and individual committee members may escape personal social censure for these, especially if their role in the decision is kept secret. Frequently, committee members hide behind the cloak of anonymity. The prevalence of socio-economic decision-making by committees has probably increased in relative importance with the development of nation states and corporate capitalism.

Economic decisions made by individuals and entities in market systems often have remote social and environmental consequences. If these consequences are negative, decision-makers may fail to take their consequences into account for a variety of reasons apart from possibly not knowing about them. For example, decision-makers may be beyond the geographical reach of those negatively affected and, as a result, escape their wrath and the local social disapproval of those disadvantaged by these decisions.

The Tyranny of Large Numbers and Failure to Exercise Moral Responsibility

In contrast to the earliest of human societies, today most individuals and economic entities have a minimal impact on collective economic outcomes. For example, purchases or sales of shares in companies by an individual or a few entities have little or no impact on the price of these shares, in most cases. Therefore, shareholders having minority holdings in companies may feel that they have little or no influence on the behaviour of large companies in which they invest. Consequently, they may continue to invest in companies for financial gain even though they morally disapprove of their behaviour or the types of commodities they produce or sell.

Anticipated Competitive Behaviour and Failure to Exercise Moral Responsibility

Closely related reasons for failure to exercise moral responsibility in economic affairs in market systems are as follows:

- if competitors do not act in a socially responsible way, competing firms may be forced to do likewise in order to survive; and
- if a profitable investment opportunity having negative social worth is ignored, other investors may undertake the investment; in that case, the investment will be made anyway.

Several aspects of the above-mentioned issues are discussed in Tisdell (1990, Ch. 2).

Modern Economic Systems as Superorganisms

Given the above observations, it can be argued that modern societies based on market systems operate like single superorganisms (see, for example, Gowdy and Krall, 2013; 2014). This implies that the nature of the behaviour and actions of individuals is largely determined by the structure of the socio-economic system itself. Societies can become so enmeshed in such systems that it is extremely difficult to control or alter their trajectory even though failure to do so may result in the complete collapse of the society, for example as a result of an environmental disaster caused by the actions of a society itself.

5.5 THE IMPACT OF CHANGING ECONOMIC SYSTEMS ON SOCIAL STRUCTURES, SOCIAL EMBEDDING AND ULTRASOCIALITY

The relationship between economic systems and the evolution of social structures is an important development issue. It is also relevant to considering changes in the degree and nature of social embedding as well as for assessing the proposition that modern societies have become as ultrasocial in their organization as some communal species of ants and bees. Consider each of these matters in turn.

The Influence of Changing Economic Systems on Social Structures

Taking into account the historical record, there can be little doubt that alterations in economic systems (changes in the fundamental methods of economic production in societies) result in significant changes in social structures, in social values, and in the degree of inequality in societies as well as the extent to which social embedding occurs. Social structures associated with the earliest human societies, which depended on hunting and gathering for their livelihood, differed considerably (on the whole) from the types of societies that emerged following the Agricultural Revolution. Many socio-economic changes occurred as a result of the development of agriculture. For example, in general, economic inequality increased and political and economic power became more concentrated. Let us consider relevant aspects of the Agricultural Revolution.

Economics and environmental change

The Agricultural Revolution and Social and Economic Change

The Agricultural Revolution began at different times in different parts of
the world and in many cases it happened independently, for example, in
the Middle East, the Americas, China and New Guinea (Renfrew, 2007).
In several countries where agriculture began, it eventually resulted in a
storable and transportable food surplus. This enabled a small elite class
to emerge. This elite was able to appropriate the agricultural surplus
and dominate other members of their community. The typical pattern
of development is illustrated in Figure 5.1. This is not to suggest that
inequality was completely absent in pre-farming communities (Pringle,
2014). They were diverse in their socio-economic structures (Svizzero and
Tisdell, 2016). However, the prevalence and the degree of socio-economic
inequality increased markedly following the commencement of agriculture.
Armstrong (2014, p. 11) states rather bluntly:

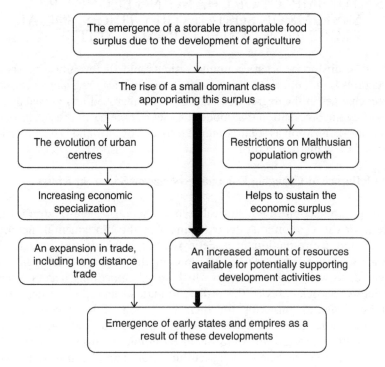

*Figure 5.1   A schematic representation of the 'typical' pattern of
development in ancient times following the commencement of
agriculture*

In the empires of the Middle East, China, India and Europe, an elite group comprising not more than 2 per cent of the population, with the help of a small band of retainers, systematically robbed the masses of the produce they had grown in order to support their aristocratic lifestyle. Yet, social historians argue, that without this iniquitous arrangement, human beings would probably have never advanced beyond subsistence level, because it created a privileged class with leisure to develop civilized arts and science that made progress possible.

However, this type of economic development did little to improve the lot of the masses. Social systems usually left them at or near subsistence levels. The effect of these systems was to conserve or increase the economic surplus, which in turn benefited the dominant class who were able to appropriate this surplus. The mechanisms involved are described in Tisdell and Svizzero (Tisdell and Svizzero, 2015; see also Svizzero and Tisdell, 2014). As a result of the economic inequality which emerged, many ancient agrarian societies were able (up to a point) to escape the Malthusian population trap, that is, they were able to avoid increased population levels which would have resulted in all the available food produced being directly consumed by food producers for their survival, leaving none that could be used to foster economic development.

Whether or not the economic surplus generated societal development depended on what the ruling class did with the amount of the surplus it appropriated. If it was invested in the creation of knowledge, productive capital works and improvements in governance, it (arguably) helped to advance civilization and development. As well, it could increase the future available economic surplus (Tisdell and Svizzero, 2015). On the other hand, if the surplus was mainly used by the ruling class for their own lavish consumption, it resulted in little or no development.

Furthermore, expenditure on war or defence could swallow up the economic surplus and reduce development prospects. However, the effects of war and defence on the level of the economic surplus and on development are complex. In some cases, territorial expansion of ancient states added to their economic surplus and resulted in the areal expansion of economic growth, as well as arguably the spread of 'civilization'. In other cases, war expenditure was economically debilitating because competing parties engaged in offsetting expenditures in preparation for wars or for their defence. In every case, a potential (or actual) Paretian improvement would have been (theoretically) possible by agreements between contending parties without resort to war/defence expenditure and actual wars. In practice, there are numerous obstacles to such agreements, which include lack of trust, and uncertainty about the relative actual or the potential military power of rivals. Unfortunately, these obstacles continue to be a problem. Moreover, the economic burden of this type of countervailing power seems to have varied historically. For example, it may have been high in medieval Europe and in China's Warring States period.

The Industrial Revolution and Social and Economic Change

It is not obvious that the process of the development of agriculture is best described as a revolution, given that the term 'revolution' may suggest that the process was very rapid. It had revolutionary consequences, but by modern standards it was a relatively slow evolutionary process, even though it involved faster changes in production methods than in earlier times.

The Industrial Revolution occurred in a much shorter period of time and also had revolutionary socio-economic, environmental and development consequences. It is probably more accurately described as being a revolution than the emergence of agriculture. However, in several respects, the Industrial Revolution merely magnified processes that had already become important following the development of agriculture such as the expansion of trade, growing economic specialization and increasing urbanization. For its success, it was dependent on the growth of the agricultural surplus being sufficient to support members of the rising urban population engaged in manufacturing and other activities.

Just as the Agricultural Revolution changed the pattern of economic development and the nature of social relations, so did the Industrial Revolution. It resulted in an accelerated rate of economic growth and an associated surge in agricultural production. There was a sufficient increase in the food surplus to support a major increase in population, most of which was not ultimately engaged in agriculture. The level of global population was able to reach unprecedented levels following the Industrial Revolution. This was so despite the rate of natural population increase declining eventually to near zero (or negative) levels in countries which successfully industrialized. While the level of world population is still rising, it is expected to stabilize around 2060.

Average levels of income rose substantially in countries participating in the Industrial Revolution. Eventually, the bulk of their population no longer existed at subsistence levels or suffered from poverty. Furthermore, major decisions about economic activity and development were no longer in the hands of small, entrenched dominant classes (as they were in agrarian societies) but were now primarily determined by market forces. Economic and social status came to be increasingly decided by success within the market systems rather than by the inheritance of property and by pre-determined social privilege.

This evolving economic system (and subsequent development of it, such as the growth of the service sector and the growth of information-based activity) has resulted in a complex and a highly interdependent global socio-economic system. Individuals and societies seem to be trapped in this system. Whether or not the degree to which they are embedded in this system exceeds the extent to which agrarian societies and hunting and gathering communities were similarly shackled within their socio-economic systems is unclear. What is, however, apparent is that the potential negative consequences of embedding in the current socio-economic system are more serious for the future of human economic welfare and the state of the environment. This is because the environmental impacts of economic activity are now being registered on a global scale. It may not be an exaggeration to claim that the contemporary socio-economic system has become a global superorganism. By contrast, ancient socio-economic societies did not have a global reach.

Neo-Malthusians are concerned about the limited ability of the current socio-economic system to stem the continuing loss of natural capital and the threat this poses for sustaining economic welfare. There are also worries about the ability of the present economic system to prevent long-term unemployment and maintain job security, for example, due to the increased adoption of new information (digital) technologies. The extent of youth unemployment and widespread under employment (even among the well educated) is a major concern. These features seem to be resulting in an increase in the proportion of the poor and economically under-privileged in the population of 'developed' countries, for example, as jobs go offshore. Furthermore, on a global scale, the population problem is not yet solved: the level of global population continues to rise. Even though the level of global population is expected to reach a maximum around 2060, its absolute increase is still anticipated to be substantial.

National differences in income levels, economic and social opportunities, in the existence of peace and in the presence of law and order have resulted in a major influx of refugees and illegal immigrants to more-developed countries. This has resulted in growing social tensions. The extent to which this development is a side-effect of the current global socio-economic system is unclear.

Are Modern Societies More Ultrasocial than their Predecessors?

Scholars are divided in their opinions about whether modern market-based societies are more ultrasocial than ancient societies. Polanyi (1944) argues that ultrasociality was more marked in ancient societies (hunting and gathering and agrarian societies) than it is in market-based ones. On the other hand, Gowdy and Krall (2013; 2014; 2015) believe that ultrasociality was relatively unimportant or absent in hunter and gatherer communities but became marked once agriculture developed and even more so with the growing dominance of market systems as a means of socio-economic organization following the Industrial Revolution.

Given the multidimensional nature of social embedding, it is difficult to determine which of these propositions is correct. This is because in some respects social embedding is stronger in modern societies and less so in other respects. For example, in most contemporary market-based economies, the social and economic roles of individuals are no longer primarily determined by their status at birth (often the social role of their father) as they were in agrarian societies and, most likely, in *some* hunting and gathering communities. Nonetheless, in India, the caste system still remains a powerful influence in determining the socio-economic roles of individuals. Greater equality of opportunity for individuals (even though this is far from perfect) exists in contemporary market-based economies than in agrarian societies. Resort to physical force and intimidation as a means for filling socio-economic roles and for maintaining social dominance seems to be much rarer now than in agrarian societies and possibly in some hunting and gathering groups.

Other dimensions of social embedding include the strength and uniformity of common values and of beliefs in societies. These dimensions probably differ between current societies and they are difficult to measure. Religious beliefs, however, appear to be less important as a unifying force in many current societies than in earlier ones and this could indicate a reduction in cultural embedding. This is not to claim that social propaganda (for example, supporting the market system) is unimportant in modern societies.

While the above factors suggest that in some respects social embedding is now less marked than in agrarian societies, individuals or small groups of individuals may find it more difficult within modern societies to alter the course of socio-economic development than was so in communities consisting of small bands of hunters and gatherers. These hunters and gatherers would have been subject to less lock-in as a result of structural factors than members of contemporary societies but they might well have been subject to greater cultural constraints in deciding their future. The large numbers involved in social decision-making in market-based economies as a major impediment to rational public choice and along with other factors mentioned above (such as the 'need' for economic growth) severely constrain the trajectories of contemporary market-based societies.

Is the Analogy of Modern Societies with the Organization of Communities of Some Species of Ants and Termites Justified?

Although drawing analogies between economic and biological/ecological relationships can be rewarding, it is necessary to be especially careful in doing so because similarities can be deceptive. Gowdy and various co-authors (Gowdy and Krall, 2014; Wilson and Gowdy, 2015) have argued that the nature of ultrasociality present in the colonies of some species of ants and termites is similar to that in current market-based human societies. However, it is relevant to note that major differences occur. First, embedding in the social organization of these insects is biologically determined, whereas in the case of humans, it is primarily culturally and structurally determined, although it is possible that biological factors also have an influence. Cultural transmission may also be important (but less so) in the case of some other species of high order social mammals. Second, the time-frame in which increased ultrasociality has evolved in human societies is much shorter than in insect communities. Third, taking into account the attributes considered in the last sub-section, it can be concluded that ultrasociality is less marked in contemporary human societies than among communal insects.

Despite these observations, in many respects social embedding in contemporary human societies is more alarming than among insects. This is because much collective human behaviour affects a substantially wider geographical area of the globe and is a more serious threat to the sustainability of existing ecosystems and the future economic welfare of humankind itself. Human beings are not yet in equilibrium with natural environments whereas communal insects are probably almost in equilibrium with their environments.

5.6 CONCLUDING COMMENTS

Cultural and structural factors seem to be the two main contributors to social embedding in human societies, even though biological factors may also play a role. The extent of social embedding has altered as economic development has occurred. Because the factors resulting in social embedding are multidimensional (and interdependent), it is difficult to determine whether social embedding is now stronger in contemporary market-based societies than it was in earlier hunting and gathering societies and agrarian ones. It was argued that, compared to ancient societies, in some respects social embedding in current societies is stronger and in other respects it has become weaker. Nevertheless, the type of current social embedding is concerning because market-based societies may well have assumed the attributes of superorganisms. Some scholars claim that modern societies have therefore become highly ultrasocial and their organization and collective behaviours are very similar to that of some communal insects. This analogy should not be accepted at face value because the sources of social embedding in human societies (identified in this chapter) are fundamentally different from those which constrain insect communities. Furthermore (for reasons which have been outlined above), the adverse ramifications of social embedding of human beings for the global environmental future and for the well-being of humankind are potentially much more serious than the consequences of ultrasociality among species of social ants and termites for their future and for the global environment.

REFERENCES

Armstrong, K. (2014), *Fields of Blood: Religion and the History of Violence*, London: Bodley Head.

Gowdy, J. and L. Krall (2013), 'The ultrasocial origin of the Anthropocene', *Ecological Economics*, **95**, 137–47. DOI: 10.1016/j.ecolecon.2013.08.006.

Gowdy, J. and L. Krall (2014), 'Agriculture as a major evolutionary transition to human ultrasociality', *Journal of Bioeconomics*, **16**(2), 179–202.

Gowdy, J. and L. Krall (2015), 'The economic origins of ultrasociality', *Behavioral and Brain Sciences*, FirstView, 1–63. DOI: 10.1017/S0140525X1500059X.

Olson, M. (1965), *The Logic of Collective Action*, Cambridge, MA: Harvard University Press.

Pigou, A.C. (1938), *The Economics of Welfare*, 4th edn, London: Macmillan.

Polanyi, K. (1944), *The Great Transformation: The Political and Economic Origins of Our Time*, New York: Farrar and Rinehart.

Pringle, H. (2014), 'The ancient roots of the 1% (don't blame farming. Inequality got its start among resource-rich hunter-gatherers)', *Science*, **344**(6186), 822–5. DOI: 10.1126/science.344.6186.822.

Renfrew, C. (2007), *The Making of the Human Mind*, London: Weidenfeld and Nicolson.

Ross, L., K. Arrow, R. Cialdini, N. Diamond-Smith, J. Diamond, J. Dunne, M. Feldman, R. Horn, D. Kennedy, C. Murphy, D. Pirages, K. Smith, R. York and P. Ehrlich (2016), 'The climate change challenge and barriers to the exercise of foresight intelligence', *BioScience*, **66**(5) 363–70. DOI: 10.1093/biosci/biw025.

Svizzero, S. and C.A. Tisdell (2014), 'Inequality and wealth in ancient history: Malthus' theory reconsidered', *Economics & Sociology*, **7**(3), 223–40.

Svizzero, S. and C.A. Tisdell (2016), 'Economic evolution, diversity of societies and stages of economic development: a critique of theories applied to hunters and gatherers and their successors', *Cogent Economics & Finance*, **4**(1). DOI: 10.1080/23322039.2016.1161322.

Tisdell, C.A. (1990), *Natural Resources, Growth and Development*, New York, NY and London: Praeger.

Tisdell, C.A. (1996), *Bounded Rationality and Economic Evolution*, Cheltenham, UK and Brookfield, VT, USA: Edward Elgar Publishing.

Tisdell, C.A. (1999), *Biodiversity, Conservation and Sustainable Development: Principles and Practices with Asian Examples*, Cheltenham, UK and Northampton, MA, USA: Edward Elgar Publishing.

Tisdell, C.A. and S. Svizzero (2015), 'The Malthusian Trap and the development in pre-industrial societies: a view differing from the standard one', *Social*

Economics, Policy and Development, Working Paper No. 59, Brisbane: School of Economics, The University of Queensland.

Tisdell, C.A. and S. Svizzero (2017), 'Optimization theories of the transition from foraging to agriculture: a critical assessment and proposed alternatives', *Social Evolution & History: Studies in the Evolution of Human Societies*, (in press).

van den Bergh, J.C.J.M. and J.M. Gowdy (2009), 'A group selection perspective on economic behavior, institutions and organizations', *Journal of Economic Behavior & Organization*, **72**(1), 1–20. DOI: 10.1016/j.jebo.2009.04.017.

Weber, E.U. and P.C. Stern (2011), 'Public understanding of climate change in the United States', *American Psychologist*, **66**(4), 315–28.

Weber, M. (1947), *Theory of Social and Economic Organization*, A.M. Henderson and Talcott Parsons [trans.], New York: Oxford University Press.

Weisdorf, J.L. (2005), 'From foraging to farming: explaining the Neolithic Revolution', *Journal of Economic Surveys*, **19**, 561–86.

Wilson, D.S. and J.M. Gowdy (2015), 'Human ultrasociality and the invisible hand: foundational developments in evolutionary science alter a foundational concept in economics', *Journal of Bioeconomics*, **17**(1), 37–52. DOI: 10.1007/s10818-014-9192-x.

6. Consumers' sovereignty – significant failures: why consumers' demands for environmental, human and animal protection are often unmet

6.1 INTRODUCTION

The view is frequently expressed in the economic literature that consumers should have freedom of choice in purchasing commodities. This proposition is usually supported by two claims, namely freedom of choice is consistent with libertarian values and it results in consumers' sovereignty. The latter in turn is deemed to result in supplies of commodities (in a competitive market system) being closely geared to the demands of consumers. A further contention, for example, strongly stated by Hayek (1948), is that consumers only need to have very little knowledge in a competitive market system to effectively signal their demands, and that the whole system economizes on the amount of information required by market participants to ensure its efficient operation in satisfying human desires. This chapter highlights the importance of qualifying these extreme claims and considers the social desirability of restricting freedom of choice by consumers in particular circumstances.

First, the basis of the idea that consumers' sovereignty results in an ideal economic outcome is outlined and then fundamental limitations of this view are stated. Secondly, consideration is given to the effects on the efficiency of market operations of asymmetry of information of market participants and their uncertainty about the attributes of commodities. Measures are then outlined and examined which might be taken by consumers, businesses and governments to address market failures which can ensue. Next the economics of food safety standards is given particular attention because consumers in contemporary societies depend heavily on purchased food to survive and mainly rely on others to ensure its safety. Today, consumers not only have concerns about the safety of their purchases but also many want to take account of the impact of the production of their purchases on the state of the environment, social and animal welfare. It can be difficult for buyers to obtain and process knowledge about these attributes (especially given the lengthening and growing complexity of product chains) and this can result in market failure. Consequently, this topic is discussed. For illustrative purposes, particular attention is given to the certification of desirable environmental and social attributes of the sourcing of wood-based products. In relation to animal welfare, special consideration is given to free-range eggs. Before concluding, adverse externalities arising from the final acts of consuming some commodities and their detrimental effects on the well-being of their buyers are examined. Alcohol use is considered as an example. Consequently, significant social issues are discussed which have received growing economic attention in recent times but which have been ignored in traditional neoclassical economics.

6.2 THE BASIS OF THE IDEA THAT CONSUMERS' SOVEREIGNTY RESULTS IN AN IDEAL ECONOMIC OUTCOME IN A COMPETITIVE MARKET SYSTEM

The idea that a competitive market system leads to an ideal economic outcome and responds well to what consumers desire was first well expressed by Adam Smith (1776 [1910]). In 1776, Adam Smith expressed two radical theses, namely that the purpose of economic activity should be to maximize the welfare of consumers; and that if individual participants in a competitive market system followed their own self-interest, this would result in this ideal economic outcome. Subsequently, these theses were provided with stronger logical/mathematical foundations by neoclassical economists, for example, Pareto (1927). This added to the persuasiveness of the thesis that a free competitive market system would provide maximum possible benefits to consumers.

Hayek (1948) also emphasized that the decentralized market system is efficient because participants in the system only have to make use of a minimal amount of information to ensure that the system satisfies consumers' demands to the maximum extent possible given the limited availability of resources. Basically, Hayek assumes that all consumers need to know in order to maximize their satisfaction are the prices of the commodities they may be interested in purchasing. This assumes that consumers are extremely knowledgeable about the attributes or qualities of the commodities they are interested in purchasing. Moreover, it is implicitly assumed that consumers are able to obtain this knowledge without any impediment or cost to themselves. This unbounded rationality assumption is an unreasonable generalization given the nature of contemporary economic systems. Furthermore, Hayek supposes that it is only the final utility consumers personally derive from their consumption of commodities that concerns them. The latter implies, for example, that consumers have no qualms about any adverse environmental externalities or social effects of their buying decisions. Nor are they worried about the consequences of these decisions for animal welfare, and their implications for the survival of species. They would need knowledge about these effects if they were concerned by them. All of the above-mentioned assumptions are contentious for reasons that will now be considered.

6.3 ASYMMETRY OF KNOWLEDGE, UNCERTAINTY, MARKET FAILURE AND CONSUMER PROTECTION

Akerlof (1970) has pointed out that the knowledge of buyers about the attributes of commodities is often poorer than that of suppliers and that this can provide an economic incentive for suppliers to pass off inferior commodities as superior ones, if it is cheaper to supply commodities with inferior characteristics. While some suppliers are dishonest, others may not be diligent in ensuring the quality of their product rather than being dishonest. In either case, the effects on the market can be the same, namely:

- only inferior products are supplied to the market;
- the quality of the products supplied to the market may become increasingly uncertain and variable; and as a result
- the demand for the marketed products may fall substantially or disappear completely.

If any of these effects arise, a social economic loss occurs because commodities are unavailable with specific qualities which consumers are willing to pay for and which some suppliers would find it profitable to provide in the absence of cheating by or lack of diligence by other suppliers.

Market participants and governments can take steps to counteract the above-mentioned possibilities. However, these all have costs and differ in their cost-effectiveness. Because of these costs, economic well-being is lower than it would have been had all suppliers been honest or if all had exercised due diligence in the quality control of their market supplies.

This raises the question of what the appropriate policy responses are to the possibilities outlined above. The choice of policies is likely to be influenced by economic considerations but may not be determined entirely by these. For example, libertarians may be wary of government intervention in the market system and reject policies that are economically defensible and protective of consumers.

The Responsibility of Different Entities for Consumer Protection

At one end of the spectrum, buyers may be entirely responsible for protecting themselves against the purchase of defective commodities or those with attributes they do not want. This is summed up in the legal adage 'let the buyer beware'. In order to protect themselves in this situation, consumers may search for more information than otherwise before deciding to buy, for example, seeking information from previous buyers, consumer organizations and other third parties, or checking prospective purchases very carefully. However, in many cases, this will not provide the consumer with complete relevant information. Often there are limits to the extent to which uncertainty about product quality *can be* reduced by the buyer prior to purchase as well as the economics of the buyer doing so.

The search behaviour of the consumer can be modelled as a cost–benefit problem. In that case, the higher is the marginal *net* economic benefit from collecting information, the greater is likely to be the buyer's effort in collecting information prior to purchase. This type of modelling is amenable to empirical investigation. For example, one might expect consumers to be more careful in their purchase of expensive consumer durables than less expensive ones (Katona, 1960, p. 148), especially if the former will have low or zero resale value if defective and as a result of the purchase, the consumer suffers a substantial financial loss.

Note that the cost to consumers of obtaining information about the characteristics of commodities is a market transaction cost. Even if uncertainty about the attributes of products offered for sale does not result in market collapse, these costs reduce the potential economic surplus consumers obtain from their purchases, and this reduction increases as these costs become more onerous.

Individual sellers (or cooperating groups of them) may find it profitable to reduce these market transaction costs voluntarily. Ways in which they can do this include by:

- the voluntary disclosure of information about the attributes of their products;
- giving guarantees about the quality of their product with offers to replace items found to be unsatisfactory, refund payments, repair products found to be defective and so on; and

- obtaining certification by a trusted third party (that is, trusted by buyers) that a product meets certain standards or has particular qualities, for instance, that a food product is organic or various animal products (such as eggs) are obtained from free-range animals.

The value of the above measures depends upon their authenticity. In the case of the provision of information by sellers, the consumer will be concerned about the accuracy and the extent to which the information disclosed is relevant to the consumer's choice. The seller may only provide information about the positive features of the product. Information supplied by sellers is likely to be biased unless they are required by law to mention specified possible negative features of a product or they increase their legal liability by not doing so.

The worth of guarantees depends on the ability and the willingness of the guarantor to honour the guarantee and the nature of the conditions placed on the guarantee. The guarantee will be of little value if it is costly to have it honoured or if the guarantor goes out of business during the guarantee period.

The effectiveness of all these measures depends on the trust buyers have in sellers. Some firms, for example, may have a reputation for only selling quality products and can be generous in permitting return of products for replacement or refund if they are found by the buyer to be unsatisfactory. If they are retailers, they may in turn be able to claim compensation from their suppliers in the product chain. Trust is therefore a valuable asset for sellers of products. Accountants recognize this as an intangible asset normally specified as goodwill. It increases the resale value of a business or of the brand of a product.

Governments are the third set of important participants in consumer protection. They may protect consumers by one or more of the following methods:

- monitoring commodities or their methods of production for compliance with quality standards;
- establishing legal frameworks which enable market participants to take legal action against sellers who fail to comply with relevant regulations or laws, for example, sell injurious products or make false claims about their products; and/or
- prosecuting suppliers of commodities that fail to comply with prescribed standards, or for which false claims are made by sellers.

Government inspectors, for example, may inspect imported food to determine if it meets health standards of the importing country. However, this can be costly and therefore usually only samples are tested. Consequently, perfect quality control is not achieved. Legal action by individuals to obtain redress for the supply of defective or injurious commodities can be costly and the outcome uncertain. This is a deterrent to legal action by a consumer, especially if the defendants are in a superior financial position to defend claims. However, if a class action is possible, this will facilitate prosecution.

Determining the most economic set of measures for government intervention in managing the quality standards of commodities is complex. Whether or not government intervention is justified and the pattern it should take is liable to depend on the nature of the particular commodity and the extent to which market failure occurs. Some issues involving food safety standards are discussed below. A related issue of interest is to examine the historical evolution of government measures and legal changes designed to protect consumers.

In many legal systems, sellers are legally required to exercise a duty of care in ensuring that the commodities they sell are not defective or injurious to buyers. If they fail to do so, they are usually legally required to compensate buyers and may also be fined. What constitutes a socially (legally) adequate standard of care is open to interpretation. In higher-income countries, a higher standard may be required than in lower-income countries, and the standard may increase as economic development occurs and new techniques for quality control become available.

In some instances, the seller's legal liability may be reduced if the seller makes the buyer aware of possible defects in a product, its actual or potential side-effects or qualities that are not guaranteed or claimed. Nevertheless, the seller cannot, as a rule, be absolved from all legal responsibility for the sale of defective, dangerous or shoddy products by statements of this nature.

In higher-income countries in particular, the frequency of the strict liability of sellers for the adverse effects of the uses of their commodities appears to have increased. Not only may the producer of a drug be legally liable for the adverse health effects of a marketed drug which are known to the supplier but not revealed to the buyer, but may also be legally liable for (major) adverse consequences of the drug which were unknown to the producer. This tends to delay the release of new drugs and adds to research costs required to determine the range of possible adverse consequences of new drugs. While the delay can have economic benefits, in some cases, the cost of the delay may be excessive from a social point of view. The costs and legal liabilities involved in the development and marketing of new drugs results in the pharmaceutical industry being dominated by a few large multinational companies. This is, for example, because they increase the absolute costs of entry of a new business to the industry and can increase the company's need to achieve economies of scale or scope in order to be economically viable.

In most countries, the legal sale of several categories of pharmaceuticals requires the approval of government authorities, for example, the Food and Drug Administration in the United States. While this is intended mainly to protect consumers, it coincidentally may also limit the legal liability of suppliers of registered drugs if these have adverse unpredicted consequences. In some instances, sellers welcome government approval for the sale of goods when it limits their legal liability for damages to buyers (Tisdell, 1983; 1993, Ch. 5).

Note that no consumer protection by governments would be economic or necessary if consumers had the degree of knowledge about the quality of products assumed in neoclassical economic theory and by Hayek (1948).The bounded rationality of buyers (and in some cases, of sellers) has important implications for well-being and public policy and should not be ignored. To reinforce this point, consider economic aspects of food safety standards.

6.4   ECONOMIC ASPECTS OF FOOD SAFETY STANDARDS

In modern economies, virtually all individuals depend on food purchases in order to survive. Most are concerned about how safe this food is to eat and about its consequences for their health. The buyer is frequently unable to determine by inspecting food whether it is safe to eat and is often igno-rant about the ingredients of purchased food and the health consequences of the consumption of food of various types. The paucity of consumers' knowledge about the consequences for their health of their food purchases poses a challenging social problem and raises significant economic policy issues.

These issues include the determination of economically optimal food standards, the economics of enforcing these standards, the economics of differing standards (for example, to satisfy the differing requirements of higher- and lower-income countries), the economics of providing informa-tion to consumers about the attributes of food and the impact of product chains on tracing defects in the quality of food. Consider these factors, bearing in mind that consumers in modern economies rely heavily on the policies of others to ensure that their food purchases do not result in their ill-health or in extreme cases, cause their death.

Economics and Optimal Food Standards

Several different approaches to determining food safety standards exist (Traill and Koenig, 2010), including various economic methods. At a very general level, the economically optimal level of food safety is that for which the marginal social cost of achieving it equals the marginal social benefit obtained. As explained by Traill and Koenig (2010, p. 1612), the marginal social benefit curve of achieving greater food safety can be expected to be downward sloping whereas the corresponding social cost curve is likely to have a positive slope.

Identifying all social costs and benefits involved and measuring these can be a formidable and controversial task. This will not be attempted here. Rather attention will be focused on the net benefits of food safety to buyers.

Two basic economic procedures exist for determining the value or benefit to consumers of food safety. These are: (1) by eliciting their willingness to pay for greater food safety by means of stated preference methods (or alternatively, by revealed preference methods); or (2) by determining the reduction in health costs that consumers can be expected to experience as a result of greater food safety. These methods can provide an indication of consumer demand for different levels of food safety but none are flawless.

Hamilton et al. (2003) adopted a stated preference approach to estimate the demand for pesticide-free food based on a Californian sample of respondents, and McCluskey et al. (2005) did this to predict the willingness to pay of a sample of Japanese consumers for beef certified as being free from bovine spongiform encephalopathy (BSE), 'mad cow disease'. While revealed preference methods may also be used (sometimes) to estimate the demand for food satisfying different safety standards (for instance, hedonic pricing methods), the observed relationships are often confounded by influences on the price of the products involved other than food safety considerations. For example, some consumers are prepared to pay a premium for organic food not only because they believe it to be healthier than non-organic food but because they consider it to be environmentally more friendly than the supply of non-organic food. The same difficulty can arise in valuing food safety using stated preference methods, for example, the absence of pesticide residues in food (Hamilton et al., 2003). Nevertheless, not all measures to improve food safety have adverse environmental consequences. For example, more rigorous inspection of meat for pathogens and greater hygiene in abattoirs are unlikely to have any major adverse environmental effects.

Another approach to evaluating the economic benefits to consumers of improved food safety standards concentrates on estimating the costs of illness avoided as a result of these standards. Its components usually consist of medical costs and loss of income avoided and an economic allowance for escaping premature death. Traill and Koenig (2010, p. 1615) provide a US Department of Agriculture (USDA) application of this approach to *Salmonella* poisoning. They point out some of its limitations, for instance, that it does not allow for the pain and anxiety caused to patients by food-borne diseases and the difficulty of placing defensible economic values on premature death.

It should also be noted that willingness to pay approaches to food safety are subject to several limitations. Consumers are often ill-informed about the health consequences of eating different types of food and the safety of doing so. This affects both their stated and revealed preferences (cf. Tisdell and Wilson, 2012, Ch. 4). Willingness to pay statements, such as the maximum amounts consumers say they would pay for an increase in food safety, may be unreliable because of 'cheap talk' resulting in statements not being backed up by corresponding actions. Again, in the case of stated preferences, individuals have to be interviewed, and they can only *imagine* what they might do and their imagination is not always a reliable guide to their actions.

Income Levels and the Demand for Food Safety

Despite the above difficulties, economic modelling and empirical results suggest that the economic benefits from increased food safety tend to rise with the income levels of consumers. The components of the cost-avoidance method support this hypothesis. Income forgone by those in higher-income brackets as a result of illness is greater than by those earning less, and measures of economic value of their avoidance of premature death are usually higher. Probably also, those on higher incomes are prepared to spend more on medical expenses. This complicates the economic assessment because as a result of their expenditure, illness of those who are well off may be cured more quickly and it may be less likely to result in premature death than in the case of the poor.

There is empirical evidence that the willingness to pay for higher food safety standards rises with income levels; see, for example, Hamilton et al. (2003, p. 808). Higher food safety standards are a type of insurance. Australian empirical evidence indicates that lower-income groups are less likely to insure against ill-health than higher-income groups (Denniss, 2005). In the United States in 2014, low- or moderate-income families were less likely to have private insurance than higher-income families (Kaiser Foundation, 2015, p. 5).

The above suggests that in lower-income countries, it can be economically optimal to have lower food safety standards than in higher-income countries. However, if greater corruption or less diligence occurs in maintaining standards in lower-income countries, this can result in actual food standards being lower than is economically optimal. Furthermore, other things held constant, if the costs of attaining food safety standards are higher in lower-income countries than in higher-income countries, this further reduces economically optimal food standards in low-income countries.

Different Food Standards for Different Markets

It is not uncommon for nations to determine different food safety standards for different markets either by government intervention or as a result of private business decisions. The higher standard may be demanded by the governments of importing countries or may be adopted voluntarily to access those overseas markets where there is greater demand for food safety than at home. Within a country food may also be supplied to satisfy differences in demand for its safety either as a result of government certification schemes or due to private business initiatives.

If consumers do not trust the safety of local food products, they may try to source substitutes from trusted foreign sources. For example, contamination of milk powder prepared for babies and infants in China (in the second decade of this century) resulted in a substantial increase in Chinese demand for this powder from New Zealand and Australia, presumably mostly by Chinese on higher incomes.

Note also that requiring suppliers in all nations in a common market, such as the EU, to meet uniform minimum health safety standards can be economically disadvantageous to lower-income nations. Depending on how stringent these standards are in a common market, their economic disbenefits to lower-income nations may outweigh their benefits.

Product Chains, Traceability and Food Safety Standards

The extension of markets and associated processes of economic globalization appear, on the whole, to have increased the length of food product chains and the diversity of the geographic sources of the components of many foods. It is not uncommon to find that processed foods are made from a combination of local and imported ingredients or that a business sells food items sourced from varied geographical areas. Increased length, diversity and complexity of product chains can increase the difficulty of tracing the source of breaches in food safety standards and add to the costs and risks to those at the end of a product chain (or near its end) of meeting their product safety standards. This tends to increase market concentration at different stages of the product chain when this concentration lowers the costs of tracing deficiencies in food quality (Hammoudi et al., 2009, p. 471). Larger-sized businesses may be in a superior position to smaller-sized ones to ensure that their supplies comply with health safety standards.

Hammoudi et al. (2009, p. 472) point out that 'public and private standards may marginalize small producers'. This is not only because large producers are better placed to satisfy higher food standards (Dolan and Humphrey, 2000; Van der Meer, 2006) but because major buyers of food items often find it less costly (and use their market power) to ensure standards are met by contracting with large-sized firms selling food items to them rather than doing so with small ones (Fulpani, 2006; Giraud-Héraud et al., 2008).

6.5 OPPORTUNITY COSTS, SUPPLY RESPONSES, AND THE INCIDENCE OF THE COSTS OF MEETING FOOD SAFETY STANDARDS

It is important that opportunity costs be adequately taken into account when assessing the economic consequences for health of food production methods, such as the opportunity costs of not using pesticides and the addition of chemicals to preserve food. Another aspect of interest is who effectively pays for measures to increase food safety? In other words, what is the incidence of the costs involved in meeting food standards?

The Need to Carefully Consider Alternatives

Many methods of supplying food have both negative and beneficial effects on health. A biased economic assessment of their health effects occurs if only their adverse consequences are considered. In some instances, chemical preservatives added to food or the chemicals used to clean equipment for producing food may have some ill effects on health. On the other hand, the alternative to using these may involve greater health risks because the chances of food being contaminated by harmful bacteria are likely to increase and food supplies may be reduced because of spoilage. Therefore, the reduced costs of ill health due to food-borne diseases resulting from the use of chemical preservatives may outweigh the cost of illness caused by their use. In addition, food availability may be increased by the use of these preservatives, thereby raising nutrition. This may be important for social groups experiencing food scarcity. Health costs can be lower for the use of chemical preservatives and methods of food production having some negative health effects compared to their absence. The results depend on particular cases, the nature of the chemicals involved and so on.

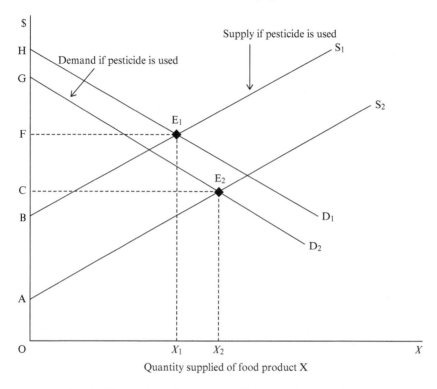

Figure 6.1 An illustration of a case in which a food production method elevates health risks to consumers but nevertheless increases their economic welfare and the total economic surplus

Employing the willingness to pay approach (and assuming that consumers are only interested in their own health), Figure 6.1 illustrates a case in which a food production method using chemicals (for example, a pesticide) has some negative effect on health but increases economic welfare due to its significant positive impact on food supply. Using partial equilibrium analysis, in the absence of the use of a pesticide for growing a food crop, X, it is supposed that the supply schedule is BS_1 and the demand schedule is HD_1. The market is in equilibrium at E_1. As a result, it is posited that yields of X increase and that its costs of production are reduced. Consequently, the supply schedule shifts to AS_2. Some reduction in the demand for X is also supposed since the use of the pesticide elevates health risks and therefore the demand schedule shifts to GD_2. A new market equilibrium is established at E_2 when the pesticide is used.

Given the parallel shifts in supply and demand schedules shown in Figure 6.1 and that the shift in the demand schedule is smaller than that of the supply schedule, both consumers' surplus and producers' surplus increase as a result of the use of the pesticide. The area of triangle CE_2G exceeds that of triangle FE_1H because both the base and height of the former triangle exceed that of the latter. Hence, consumers' surplus increases. Similarly, producers' surplus increases because the area of triangle AE_2C exceeds that of triangle BE_1F for the same reason.

This is a simplified case because both changes in the intercept values and the slopes of the demand and supply schedules can occur and affect the distribution of economic benefits. Of course, even in the simplified case illustrated in Figure 6.1, if the downward movement of the demand schedule as a result of pesticide use exceeds that of the supply schedule, both the economic surplus of consumers and that of producers declines.

The Incidence of the Costs of Satisfying Food Safety Standards

Determining who pays for the costs of satisfying food standards can be complicated. If the costs are initially met by suppliers, they are likely (to some extent) to be passed on to the consumers, depending on the relative elasticities of demand and supply curves. However, while this case has similarities with the incidence of taxes on production, there are differences because a change in mandated food safety standards usually results in a shift in the demand curves for the products involved. Figure 6.2 provides a simple example of the difference. In Figure 6.2, the supply schedule AS_1 applies prior to an increase in the food safety standard for a product, X, and CD_1 represents the corresponding perfectly elastic demand schedule. The market is in equilibrium at E_1 initially.

Suppose that an increased food standard is mandated for product X and that this raises the per unit cost to producers of supplying it by the amount AB. The supply schedule moves up from AS_1 to BS_2. If the demand for the product is unchanged, a new market equilibrium is established at E_2. The economic burden of satisfying the standard then falls completely on producers and the equilibrium price of the product is unchanged. This case is equivalent to a tax on production. However, one would expect an improved food safety standard for product X to raise its demand schedule. If this happens, the market equilibrium price of the product rises. If, for simplicity, the demand schedule shifts up by a constant equivalent to CF, producers will be exactly compensated for meeting the new standard because a new market equilibrium is established at E_3. The increased price of the product exactly compensates for the extra cost of meeting the increased safety standard. Partial compensation will occur if the demand schedule moves up by less than CF and producers will be more than compensated if this movement exceeds CF. Nevertheless, in cases where the increased

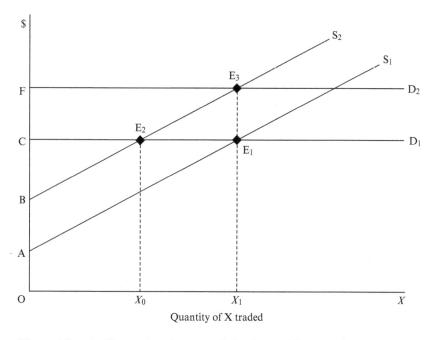

Figure 6.2   An illustration that even if the demand for a product is perfectly elastic, the economic burden of meeting an increased food safety standard may not fall entirely on producers

market price of meeting a higher food safety standard compensates or more than compensates producers for their extra costs of satisfying higher food safety standards, producers may still complain of the cost burden because they fail to take account of market adjustments.

6.6   CONSUMERS' CONCERNS ABOUT THE EFFECTS ON THE STATE OF THE ENVIRONMENT, SOCIAL AND ANIMAL WELFARE CAUSED BY THE PRODUCTION OF THEIR PURCHASES

Are consumers' demand for commodities influenced (and if so, to what extent) by the consequences of the production of commodities for the state of the environment, social well-being and animal welfare? If an instrumentalist point of view is adopted, one might argue that this is unlikely because each individual consumer's purchase of a commodity constitutes a miniscule amount of the demand for it. Therefore, because of the large number of independent consumers of most commodities, each will conclude in a competitive market that their individual purchasing actions will have no effect on environmental, social and animal welfare factors associated with the supply of these commodities.

On the other hand, if one considers Kantian-like ethics, some individuals may refuse to buy commodities the production of which they believe has unfavourable environmental social and ethical welfare effects, even if they know that their private choice is virtually inconsequential for the aggregate supply of these. In doing this, they are sometimes said to purchase moral satisfaction. More importantly, if enough demand exists for commodities that are produced in a preferred environmental, social or animal welfare manner, private businesses have an economic incentive to supply these if this is profitable, for example, if consumers are willing to pay a price premium to cover (or more than cover) any extra cost involved in satisfying these demands. However, consumers who demand such attributes must be made aware of the presence of these attributes and need to be convinced of the relevant claims. Asymmetry of information is important in this case and tends to increase as product chains become longer and more complex. Inspection of a final product supplies little or no information to the buyer about the environmental consequences and welfare effects on other sentient beings of its supply, for example, in the absence of relevant and believable labelling by the seller. These products have been described as credence goods.

Depending on the nature of the product and the demand for it, several examples are available of businesses and charities which try to satisfy demands for one or more of the above-mentioned attributes. Some businesses may do this to satisfy the demand of consumers as well as to deflect or prevent possible social criticism. Furthermore, some business managers may wish to purchase moral satisfaction by taking ethical considerations into account. The extent to which they can do this depends on their economic scope for discretionary behaviour. Some shareholders may also have an interest in the ethical behaviour of companies in which they invest.

Examples of Business Practices Designed to Satisfy the Social Values of Consumers and Other Stakeholders in the Supply of Commodities

Certification by independent parties (having a reputation for trustworthiness and reliability) is one means by which businesses can obtain recognition for socially approved behaviours in supplying commodities. For example, a business supplying products made from wood (for example, furniture or paper) may decide that it will supply products which meet the criteria of the Forest Stewardship Council (FSC) and apply for certification. Certification entitles the business to use the logo of the FSC on its supply of complying products.

FSC International (2015) states that its vision is that 'the world's forests meet the social, ecological and economic rights and needs of the present generation without compromising those of future generations' and that it aims to do this by promoting 'environmentally appropriate, socially beneficial, and economically viable management of the world's forests'. These criteria are similar to some which have been suggested for achieving sustainable development (Tisdell, 1994). Although these goals appear to be socially very desirable, careful consideration suggests that they are of limited operational value because they are very general in nature and do not adequately take account of unavoidable trade-offs in the fulfilment of these goals. To assess the effects of FSC certification, one has to consider empirically how accreditation is given. This cannot be done here but some observations are in order.

FSC International (2015) states:

> Economically viable forest management means that forest operations are structured and managed so as to be sufficiently profitable, without generating financial profit at the expense of the forest resource, the ecosystem, or affected communities. The tension between the need to generate adequate financial returns and the principles of responsible forest operations can be reduced through efforts to market the full range of forest products and services for their full value.

Despite the above claim, it seems highly unlikely that commercial forestry will have no significant effect on pre-existing ecosystems.

FSC has different categories of certification. One classification is 'mixed sources'. This covers production from a combination of FSC certified and 'controlled' wood. The controlled category excludes the use of some types of wood, for example, wood harvested in forests being converted to plantations or non-forest use. The criteria for what constitutes controlled wood are more liberal than in the case of certified wood and monitoring of the use of this type of wood appears to be much more lenient.

FSC International (no date) states: 'Controlled wood is a material that can be mixed with certified material during manufacturing FSC mix products. This has enabled manufacturers to manage low and fluctuating supplies of FSC certified products while creating demand for FSC certified wood'. However, there appears to be no empirical evidence to demonstrate that FSC mixed sources certification has increased the demand for FSC certified wood. Nevertheless, FSC International (no date) maintains that FSC mixed product categories (which enable suppliers to use the FSC logos) have increased the number of manufacturers seeking its certification of their product and make the FSC logo more visible to consumers.

The extension of the rights of suppliers of products to use the FSC logo weakens the environmental and social implications of the use of its logo. The reasons for this extension may include the following:

- FSC may be responding to the demands of its clients, that is, suppliers of wood products seeking certification.
- It may be a reaction to the practices of its competitors – more than one not-for-profit organization provides eco-certification of wooden products.
- It may be motivated by the FSC wanting to increase its market share or at least not see its revenue from certification decline.

The behavioural motivations of NGOs appear to be mixed. On the one hand, they usually have altruistic intent but also they are motivated to survive economically and in most cases, grow. Moreover, they are normally not free of competition from other NGOs for funding or pressure from their 'clients'.

Findings conflict about the willingness of buyers to pay for eco-certification and for the less restrictive types of certification arranged by the FSC through its accredited certifiers. Surveys relying on contingent valuation questions conclude that consumers are willing to pay a premium for certified forest products (Cai and Aguilar, 2013; Elliott, 2014; Jenson and Jakus, 2003). On the other hand, Anderson (2003), as a result of directly observing the actual behaviour of a group of buyers of timber products, concluded that most were not willing to pay a premium for certification. Because consumer ignorance about the nature of certification is widespread, this suggests that it may have little effect on buyer behaviour. Furthermore, its positive impact on purchases (when a price premium is demanded) is definitely less than indicated by contingent valuation studies because these studies provide most respondents with extra information and the usual limitations of contingent valuation analysis apply. Elliott (2014) found by surveying a small sample of buyers of paper that only 3 per cent said that they were well informed about the standards for eco-certification of this paper. Forty-eight per cent responded that they knew nothing about these standards and the remainder reported that their knowledge of these was quite limited. Aguilar and Cai (2010) and Chen et al. (2011) obtained similar results. Therefore, one should expect that predictions of purchasing behaviour based on contingent valuation studies will differ from actual behaviour.

Several well-known retail chains state that their supplies are ethically sourced and provide information about this on their websites. They include Starbucks (Starbucks Australia, no date), McDonald's Australia (no date) and The Body Shop (no date). Starbucks Global Responsibility Report 2013 (available at http://www.starbucks.com/responsibility/global-report) states 'At Starbucks, we have always aspired to be known as much for our commitment to social responsibility as we are for the quality of our coffee'. The charity, Oxfam, also operates several retail outlets and stresses its ethical ('fair trade') practices (Oxfam Australia, no date). These are just a few examples of organizations which have decided to inform the public of the ethical sourcing of their supplies. In addition, many public corporations include a section in their annual reports mentioning their socially responsible behaviours. In some cases, these may represent little more than tokenism. Furthermore, their ethical commitments are often stated in very general or intentional terms which permit considerable leeway for actual behaviour. Also they can be misleading if only positive ethical behaviours are reported. Nevertheless, the fact that many organizations are giving increasing attention to social concerns about their ethical practices is a significant development.

Animal Welfare Considerations

Particularly in higher-income countries, animal welfare concerns about the sourcing of animal-based products have intensified in recent times. As a result, the demand for products having favourable animal welfare attributes has increased. A meta-analysis (by Lagerkvist and Hess, 2011) found that the stated willingness of consumers to pay extra for food produced in a manner favourable to animal welfare tends to increase with income and decline with age. The demand for free-range eggs, for example, has been studied as an indicator of the demand for supporting animal welfare. In an early Australian study involving a survey of households, Rolfe (1999) found that 60 per cent of respondents supported a ban on intensive poultry production, 17 per cent were against this and 23 per cent were undecided. However, only 11.37 per cent of respondents said they would be willing to pay an extra $1.50 for a carton of a dozen eggs if there was a ban on battery hens. Rolfe (1999, p. 201) concludes: 'This indicates that while people may vote for a ban on battery hens when there are no explicit costs involved, the proportion who would support a ban falls substantially as costs become explicit.' This is a similar conclusion to that of Hamilton et al. (2003) in relation to pesticide use.

Parker (2013, p. 60) points out that what constitutes the category of a free-range egg in Australia is very wide and ambiguous, as it presumably is in several other countries. She maintains that major retailers determine their standards for free-range eggs and that this limits choices by consumers who are unaware of the substantial differences in criteria for what constitutes free-range eggs (Parker, 2013, Table 1). Labelling does little to inform consumers and romanticized advertising (Parker, 2013, Table 2) may in fact mislead consumers. The dominant Australian retail food chains (in Parker's opinion) support intensive free-range egg production and she claims that 'consumers cannot bring an alternative product into existence through the power of choice alone: the choices have already been constructed and constrained by actors in the production, distribution and exchange chain who bring products to retail' (Parker, 2013, p. 54).

We can conclude from this section that knowledge deficiencies remain a significant barrier to the exercise of rational choice in the case of credence goods, such as those with ethical consequences that cannot be directly observed by consumers. However, consumers do have to grapple with the possibility of information overload, given their bounded rationality. It is also possible that some consumers will be happier relying on standards that are presented in a 'sanitized' way rather than knowing about the disturbing realities of several forms of animal-based production or unethical aspects of production, including serious ecological and environmental damages. This conforms with the adage that 'ignorance is bliss'.

6.7   EXTERNALITIES ARISING FROM FINAL ACTS OF CONSUMPTION AND RESTRICTIONS ON COMMODITIES CONSUMED TO BENEFIT USERS

The previous section focused on consumers' sovereignty and the presence of environmental externalities, and the occurrence of adverse social and animal welfare effects of production processes. It must also be recognized that the final act of consuming many commodities can have adverse spillovers, that is, they may adversely affect the welfare of others. Furthermore, the consumption of some commodities can have adverse health and other deleterious personal consequences for the consumer. These usually depend upon the amount of the commodity consumed and its nature. The adverse social effects (spillovers) and negative personal consequences of consumption activities appear to show a positive association in most cases. The existence of these phenomena pose a dilemma for those who have a strong preference for economic liberalism because in the above-mentioned circumstances, freedom of choice does not promote the social good according to most communal standards.

Depending on the magnitude of the perceived social and personal damage caused by the final consumption of commodities having injurious consequences, public interventions of various kinds occur both by the state and by charitable bodies. Considerable debate exists about the (economic) effectiveness of adopted policies for controlling the consumption of deleterious commodities. Detailed analysis of alternative policies (and of this whole subject) is not possible here (due to space limitations) but some observations will help to underline its importance. Its importance appears to have increased with economic development for the following reasons:

- Deleterious commodities with negative social and personal effects are more widely available and promoted by suppliers as a result of economic development.
- Closer settlement (urbanization) has increased the scope for negative social interactions.
- Some technological changes (for example, use of automobiles capable of speed) have increased the risk of social harm from the consumption of some commodities, for example, alcohol and other drugs.

According to Rehm et al. (2009, p. 2223): 'The industrialization of production and globalization of marketing and promotion of alcohol have increased both the amount of worldwide consumption and the harm associated with it'. They provide empirical evidence of the economic cost attributable to alcohol use relying on meta-analysis. The studies taken into account basically adopt the cost of avoidance approach – they measure the health costs and social costs which could be avoided in the absence of alcohol consumption net of any beneficial effects which might occur. Rehm et al. report the personal plus social costs as a percentage of GDP (shown in brackets) of alcohol consumption for the following high-income countries: France (3.7), USA (2.7), Scotland (1.4), Canada (1.4), and for the following middle-income countries: South Korea (3.3) and Thailand (1.2). The costs of adverse social spillovers (for example, harm to others) account for the largest proportion of these costs.

Anderson et al. (2009) explore the cost-effectiveness of policies and programmes to reduce the harm caused by alcohol and find that increased price and reduced availability are relatively cost-effective in reducing alcohol consumption and its potentially harmful effects. They also state that banning alcohol advertising and measures to counter drink-driving are cost-effective measures. However, the economic and social evaluation of control is quite complicated because, for example, prohibiting the sale of alcohol or taxing it very heavily results in a new set of problems with social costs, namely illicit sale of alcohol with associated underworld

activities and problems about the safety of the illicit product. What the
appropriate social behaviour is in terms of availability, pricing and so on is
still a subject of continuing debate.

Anderson et al. (2009, p. 2243) state the following:

> Since there are substantial commercial interests involved in the promotion of
> alcohol's manufacture, distribution, pricing and sales, the alcohol industry
> has become increasingly involved in the policy arena to protect its commercial
> interests, leading to a common claim among public health professionals that
> industry is influential in setting the policy agenda, shaping the perspectives
> of legislators on policy issues, and determining the outcome of policy debates
> towards self-regulation.

This is consistent with Galbraith's thesis that large corporations and
commercial interests have a disproportionate influence on public policy
(Galbraith, 1952; see also Tisdell, 1982, p. 278). On the other hand,
Australia (despite vested business interests) has been able to adopt public
policies that have substantially reduced the smoking of tobacco, a substance well known to have serious long-term adverse health consequences
for smokers as well as others subject to inhalation of their smoke (see for
example, Galea et al., 2004). Other issues which challenge modern societies include problem-gambling and diet-related obesity. However, enough
has been said to bring attention to the issues raised to illustrate the limitations of treating economic liberalism as an absolutely ideal social goal and
placing complete reliance on consumer sovereignty to promote the 'social
good'.

6.8   CONCLUDING COMMENTS

Neoclassical economics assumes that consumers know all the relevant
characteristics of commodities they may wish to purchase, or that they
can determine these at little cost. These assumptions are far from satisfied
in modern economies, as has been demonstrated. Therefore, substantial
issues involving choices by consumers are overlooked.

Hayek (1948) proposed that all consumers need to know in order for a
competitive economy to operate efficiently (and for consumers' sovereignty
to prevail) are the prices of commodities they may wish to purchase.
Although this pleasing result (which is supportive of economic liberalism)
can be proven mathematically (given highly restricted assumptions), unfortunately the necessary assumptions for it to be satisfied are not fulfilled in
practice. This is because consumers are often uncertain, ill-informed or have
no knowledge of at least some of the significant qualities of commodities

which they may buy. Consequently, various types of market failure occur. In extreme cases, lack of trust in the quality of goods purchased can result in complete collapse of their markets or may result in only goods of inferior quality being supplied, despite adequate demand being present for the goods driven from the market. In these cases, consumers cannot rely only on their knowledge of prices to make rational choices.

As mentioned, consumers, businesses and governments can react in a variety of ways to increase consumer protection in situations where uncertainty exists about the quality of purchased commodities. It is clear from Akerlof (1970) that in many cases businesses, by adopting policies to increase buyer trust in their products, also protect their own market.

The existence of quality standards can pose a significant barrier to the entry of new firms to an industry and increase market concentration in the supply of products. Increases in the length of product chains and the variety of components obtained from these chains for use in the final products may well have exacerbated this effect.

Given the high dependence of individuals in market economies on purchases of food, the safety and quality of purchased food is of considerable social importance. Various economic measures indicate that consumers are more willing to pay for food safety (and quality) as their income rises, other things being held constant. Therefore, a legislated uniform food standard can disadvantage lower-income groups and less-developed regions. It is also worth noting that desirable food standards and preservation practices are not independent of food habits in the households of purchasers. For example, in many lower-income countries, bacterial contamination of meat in slaughter houses is often significant but most consumers do not store meat for any length of time before cooking it. This reduces the possible adverse health consequences of meat purchases because the time between slaughter of animals and the cooking of meat is short. The length of time between slaughter and meat consumption in higher-income countries is considerably longer and therefore extra measures are required to ensure the safety of meat consumption. Social and technological changes (which occur with economic development) significantly alter the demand for food safety standards.

Particularly in higher-income countries, substantial numbers of consumers are concerned about the environmental, social and animal welfare effects of the production of their purchases, and some are willing to pay to avoid those effects considered by them to be deleterious. This is supported by stated preferences and actual behaviours, for example in the case of free-range versus caged eggs. Nevertheless, studies cited in this chapter indicate that stated willingness to pay for these potentially desirable production attributes is not a perfect guide to actual behaviours. Usually, stated willingness to pay values overstates actual effective demand for these attributes. This is a common finding about stated valuation techniques. Despite this, stated values provide useful qualitative information about the demand of buyers for environmental, social and animal welfare attributes associated with production. They may also indicate whether it would be worthwhile to conduct experiments to determine the extent to which actual (rather than hypothetical) choices are influenced by knowledge of the production-related characteristics being considered.

Even when information is signalled to consumers about these qualities by certification, logos or descriptions on products, many problems remain. Consumers may be ill-informed, for example, about the qualities being signalled. They may be misled by the presence of a symbol or term used to identify products, the production of which is purported to have desirable environmental, social and animal welfare effects, for example the production of 'free-range' eggs.

Studies considered in this chapter indicate (not unexpectedly) that referendum support for governments to adopt policies to protect consumers from socially undesirable attributes of credence goods is usually stronger than their willingness to pay or actually pay for the absence of these attributes. There are many reasons why this can be so. For example, the referendum question may not make it clear that government intervention will involve extra costs and that these will have to be covered by taxpayers, and furthermore, producers may experience extra costs which (depending on comparative elasticities of supply and demand) are liable to result in higher prices for the products concerned. Nevertheless, this option can be attractive to special interest groups of consumers if these extra taxes are paid both by those who favour these measures and those who do not. In this case, beneficiaries are subsidized by non-beneficiaries for the policy action taken.

The thesis that consumers' sovereignty maximizes economic or social welfare in a competitive economy is untenable in situations where final acts of consumption generate significant negative external effects. It is also problematic in cases where individuals consume products or consume these to such an extent that their own well-being is socially impaired. These two types of situation appear frequently to be positively associated; for example, excessive alcohol consumption results in self-harm and has significant negative (net) social spillovers, according to several studies.

In the light of the above observations, there are clearly limits to the social desirability of economic liberalism in modern economies. With the expansion of markets and structural and technological changes in economic systems (for example, lengthening product chains) as a result of economic development, the types of issues raised in this chapter have grown in importance. Economics should continue to move beyond the narrow confines of neoclassical economic theory in order to address these continuing issues.

REFERENCES

Aigular, F.X. and Z. Cai (2010), 'Conjoint effect of environmental labeling, disclosure of forest of origin and price on consumer preference for wood products in the US and UK', *Ecological Economics*, **70**, 308–16.
Akerlof, G.A. (1970), 'The market for "lemons": quality uncertainty and the market mechanism', *Quarterly Journal of Economics*, **84**(3), 488–500.
Anderson, R.C. (2003), *An analysis of consumer response to environmentally certified, ecolabeled forest products*, Ph.D in Forest Products, Forestry, Oregon State University, accessed 25 July 2016 at https://ir.library.oregonstate.edu/xmlui/bitstream/handle/1957/7526/Anderson_Roy_C.pdf?sequence=1.
Anderson, P., D. Chisholm and D.C. Fuhr (2009), 'Effectiveness and cost-effectiveness of policies and programmes to reduce the harm caused by alcohol', *The Lancet*, **373**(9682), 2234–46.
Cai, Z. and F.X. Aguilar (2013), 'Meta-analysis of consumers' willingness-to-pay premiums for certified wood products', *Journal of Forest Economics*, **19**, 15–31.
Chen, J., J.L. Innes and R.A. Kozak (2011), 'An exploratory assessment of attitudes of wood products manufacturers towards forest certification', *Journal of Environmental Management*, **92**, 2984–92.
Denniss, R. (2005). 'Who benefits from private health insurance in Australia?', WP72_8. The Australian Institute, accessed 25 July 2016 at http://www.tai.org.au/documents/downloads/WP72.pdf.
Dolan, C. and J. Humphrey (2000), 'Governance and trade in fresh vegetables: the impact of UK supermarkets on the African horticultural industry', *Journal of Development Studies*, **37**, 147–76.
Elliott, J. (2014), 'An analysis of willingness to pay and reasons for purchasing certified forest products', Master's project, Nicholas School of the Environment, Duke University, accessed 21 July 2016 at http://www.unece.lsu.edu/certificate_eccos/documents/2015Mar/ce15-04.pdf.
FSC International (2015), 'Our vision and mission', accessed 21 July 2016 at https://ic.fsc.org/en/about-fsc/vision-mission.
FSC International (no date), 'Controlled wood', accessed 21 July 2016 at https://ic.fsc.org/en/certification/types-of-certification/controlled-wood-02.
Fulpani, L. (2006), 'Private voluntary standards in the food system: the perspective of major food retailers in OECD countries', *Food Policy*, **31**, 1–13.
Galbraith, J.K. (1952), *American Capitalism: The Concept of Countervailing Power*, Boston, MA: Houghton Mifflin.
Galea, S., A. Nandi and D. Vlahov (2004), 'The social epidemiology of substance use', *Epidemiologic Reviews*, **26**(1), 36–52.
Giraud-Héraud, E., A. Hammoudi, R. Hoffman and G. Soler (2008), 'Vertical relationships and safety standards in the food marketing chains', *ALISS Working Paper No. 2008-07*, Ivry-sur-Seine: Unité Recherche Alimentation et Sciences Sociales, INRA.

Hamilton, S.F., D.L. Sunding and D. Zilberman (2003), 'Public goods and the value of product quality regulations: the case of food safety', *Journal of Public Economics*, **87**, 799–817.

Hammoudi, A., R. Hoffman and Y. Surry (2009), 'Food safety standards and agri-food supply chains: an introductory overview', *European Review of Agricultural Economics*, **36**, 469–78.

Hayek, F. (1948), *Individualism and Economic Order*, Chicago, IL: Chicago University Press.

Jenson, K.L. and P.M. Jakus (2003), 'Consumers' willingness to pay for eco-certified wood products', *Economic Research Institute Study Papers No. 261*, Utah State University, accessed 21 July 2016 at http://digitalcommons.usu.edu/cgi/viewcontent.cgi?article=1260&context=eri.

Kaiser Foundation (2015), 'Key facts about the uninsured population', The Henry J. Kaiser Family Foundation, accessed 25 July 2016 at http://kff.org/uninsured/fact-sheet/key-facts-about-the-uninsured-population/.

Katona, G. (1960), *The Powerful Consumer*, New York: McGraw-Hill.

Lagerkvist, C.J. and S. Hess (2011), 'A meta-analysis of consumer willingness to pay for farm animal welfare', *European Review of Agricultural Economics*, **38**(1), 55–78.

McCluskey, J.J., K.M. Grimsrud, H. Ouchi and T.I. Wahl (2005), 'Bovine spongiform encephalopathy in Japan: consumers' food safety perceptions and the willingness to pay for tested beef', *Australian Journal of Agricultural and Resource Economics*, **49**, 197–209.

McDonald's Australia (no date), 'Macca's & the Environment', accessed 21 July 2016 at https://mcdonalds.com.au/learn/responsibility/maccas-and-the-environment/initiatives-and-trials.

Oxfam Australia (no date), 'Ethical trading', accessed 21 July 2016 at https://www.oxfam.org.au/what-we-do/ethical-trading-and-business.

Pareto, V. (1927), *Manuel d'Economie Politique*, 2nd edn, Paris: Giard.

Parker, C. (2013), 'Voting with your fork? Industrial free-range eggs and the regulatory construction of consumer choice', *The ANNALS of the American Academy of Political and Social Science*, **649**(1), 52–73.

Rehm, J., C. Mathers, S. Popova, M. Thavorncharoensap, Y. Teerawattananon and J. Patra (2009), 'Global burden of disease and injury and economic cost attributable to alcohol use and alcohol-use disorders', *The Lancet*, **373**(9682), 2223–33.

Rolfe, J. (1999), 'Ethical rules and the demand for free range eggs', *Economic Analysis and Policy*, **29**(2), 187–206.

Smith, A. (1776 [1910]), *The Wealth of Nations*, Everyman's edn (first published 1776), London: Dent and Sons.

Starbucks Australia (no date), 'Ethical sourcing', accessed 21 July 2016 at http://www.starbucks.com.au/Global-Responsibility.php.

The Body Shop (no date), 'Ethical trade', accessed 21 July 2016 at http://www.thebodyshop.com.au/our-commitment/ethical-trade.aspx#.V5a37Pl95aQ.

Tisdell, C.A. (1982), *Microeconomics of Markets*, Milton: John Wiley.

Tisdell, C.A. (1983), 'Law, economics and risk-taking', *Kyklos*, **36**(1), 3–20.

Tisdell, C.A. (1993), *Environmental Economics: Policies for Environmental Management and Sustainable Development*, Aldershot, UK and Brookfield, VT, USA: Edward Elgar Publishing.

Tisdell, C.A. (1994), 'Sustainability and sustainable development: are these concepts a help or a hindrance to economics?', *Economic Analysis and Policy*,

24(2), 133–50, reprinted in C.A. Tisdell (2003), *Ecological and Environmental Economics*, Cheltenham, UK and Northampton, MA, USA: Edward Elgar Publishing.

Tisdell, C.A. and C. Wilson (2012), *Nature-based Tourism and Conservation: New Economic Insights and Case Studies*, Cheltenham, UK and Northampton, MA, USA: Edward Elgar Publishing.

Traill, W.B. and A. Koenig (2010), 'Economic assessment of food safety standards: costs and benefits of alternative approaches', *Food Control*, **21**, 1611–19.

Van der Meer, K. (2006), 'Exclusion of small-scale farmers from coordinated supply chains', in R. Rube, M. Singerlang and H. Nijhoff (eds), *Agro-food Chains and Networks for Development*, Amsterdam: Springer, pp. 209–17.

7. Biological conservation and human-induced environmental change: contemporary socio-economic challenges

7.1 INTRODUCTION

The study of biological conservation and management is a wide one of which the quest for sustaining biodiversity is just a part. Nevertheless, in modern times, loss of biodiversity and ecosystems have become major concerns because of the scale of these losses and because they are, to a large extent, the result of human actions: mostly economic activity. These concerns are evidenced by publications such as the *Millennium Ecosystem Assessment* (2005) and *The Economics of Ecosystems and Biodiversity* (Kumar, 2010). Losses in the biodiversity of ecosystems and the variety of species are usually interrelated. Fewer ecosystems can result in fewer species existing globally or in a particular geographical area. In some cases, changes in the composition of species (for example, the introduction of exotic species) reduce the variability of ecosystems and may also reduce the variety of species existing locally as well as globally.

It is not only loss of wild biodiversity (the loss of wild species, genetic material and natural ecosystems) which is of social concern, but also losses in the human-developed genetic pool of domesticated biota, principally cultivated plants and domesticated livestock. These latter losses are primarily a result of economic development paths and they are also usually associated with alterations in ecosystems, such as agro-ecosystems.

Why are there social concerns about losses in the pool of genetic material (and its expression in living things) as well as changes in ecosystems? A basic reason is that, as a rule, genetic losses are irreversible and ecosystem losses may also be irreversible or only reversible at tremendous cost. Consequently, the danger exists of genetic resources (both natural and human developed, that is, heritage resources) being lost permanently which would have been worthwhile conserving. These losses:

- could threaten sustainable economic development;
- reduce the satisfaction that individuals experience by interacting with the living world, for example, through tourism and recreational activities; and
- are a source of feelings of 'guilt' or regret among some individuals who feel a responsibility for the conservation of living things.

The socio-economic analysis of biological conservation and management covers a wide field and is complex. Therefore, only a limited coverage of it is possible in a single chapter such as this one. This chapter is developed initially by identifying several of the different broad subjects that can be investigated in studying biological conservation and management. Then different motives are identified that can influence decisions about biological conservation and management. Subsequently, attention is given to the role of markets in determining biological conservation and management and after that to the role of non-market institutions (governments and NGOs) in doing so. The usefulness of economic valuation techniques in relation to this subject is assessed and particular attention is given to the need to take account of opportunity costs, the importance of regular biases in conservation preferences, and the difficulty of resolving social conflict about the management of biological resources. Before concluding, aspects of the following topics are discussed:

- conflicts, valuation issues and the costs of policies for conserving koalas;
- the role of wildlife rehabilitation centres in nature conservation;
- ecotourism enterprises and the conservation of species; and
- conflicts between conservationists about conserving species as illustrated by the presence of wild horses (brumbies) in the high country of Australia.

7.2 BIOLOGICAL CONSERVATION AND MANAGEMENT COVER A DIVERSE RANGE OF TOPICS AND ISSUES

Today, most studies of biological conservation and management concentrate on issues involving the preservation of biological diversity. This is understandable because there has been a large reduction in the extent of net biodiversity in modern times and well-founded concerns that this continuing loss could threaten sustainable development and human well-being. Studies of biodiversity do not as a rule simultaneously cover all living things everywhere. They are usually more partial in nature. They can, for example, vary in their spatial consideration of wild biodiversity. This may be examined at different geographical scales: local, regional or global, and the sets of species involved can vary.

Investigations of biodiversity may cover not only the existence of genetic diversity present in biota but also the diversity of assemblages of biota, for example, ecosystems both natural, human modified or constructed. Sometimes, these investigations are extended (particularly by anthropologists and sociologists) to take account of the cultural/socio-economic diversity associated with these assemblages.

Studies of biological conservation and management not only focus on the conservation of biodiversity. Many such studies are more circumscribed in their scope. They are concerned with the preservation of particular species or restricted sets of species or assemblages of biota on restricted spatial scales. Pursuing this limited type of objective may help to conserve global biodiversity but, depending on the circumstances, can also reduce it.

Ecological and associated socio-economic research may also be divided into that examining domesticated biota and wild biota as well as linkages between these. This research can be conducted at different levels of generality.

Pest control (including elimination of pests) also forms a part of socio-economic and biological studies. Depending on the circumstances, pest control can increase or reduce biological diversity globally or on a more limited spatial scale. Furthermore, it can generate considerable social conflict and controversy. Both the methods used for control of pests and the species targeted by pests can generate substantial social disagreement. This conflict can be generated by divergent social values, differences in self-interest and opposing opinions about the effectiveness of control measures. In general, human influences on biota are often an important source of social conflict, especially if public (government) intervention is involved.

Tree-clearing laws in Queensland provide an example of such conflict. In 2016, the Queensland Government tried to change the laws governing the clearing of trees on agricultural land, which had been significantly relaxed by the previous government. The stated benefits of reduced tree-clearing included greater CO_2 retention, less soil erosion (resulting, for example, in reduced sedimentation of the Great Barrier Reef) and enhanced conservation of wild biodiversity. Reduced sedimentation of the Great Barrier Reef was also seen as a way to help sustain the economic profits of the tourist industry reliant on reef resources. Furthermore, efforts to reduce tree-clearing were motivated by the possibility that UNESCO might reclassify the Great Barrier Reef Marine Park as endangered, and that it could eventually lose its World Heritage status.

On the other hand, many agriculturalists saw this proposed legislation as a threat to their livelihoods, their prospective profits and as limiting the scope for the economic development of their properties. They therefore opposed the proposed legislation. Their political lobby group, AgForce, opposed the bill whereas other lobby groups, such as the World Wildlife Fund (WWF) supported it. When the bill was introduced to the Queensland Parliament on 18 August 2016, it was defeated by just one vote. This social conflict remains unresolved. Even if it had been shown that there would have been a major aggregate net economic benefit from the passage of the bill, it still may not have been passed. Economic arguments alone are often ineffective in dealing with social conflict.

Although the Queensland Government stressed the significance of its bill for the contribution it was expected to make to the health of the Great Barrier Reef, possibly the largest area of land affected by the bill would have been river catchments in western Queensland that do not drain into this reef area. Reduction in tree-clearing would have had conservation benefits in these areas, even if these happened to be smaller than along the coast. Nevertheless, regional differentiation in controlling environmental change can be desirable (see the last chapter in this book).

7.3 DIVERSE MOTIVES AND PERCEPTIONS INFLUENCE HUMAN ATTITUDES ABOUT THE CONSERVATION AND MANAGEMENT OF BIOTA

The range of motives influencing human attitudes to the conservation and management of biota can be classified in many different ways. They can reflect self-interest, altruism, or a combination of both. They are subject to social influences as are beliefs about the relationship between human actions and their consequences for biota and, in turn, the effects of these consequences on human beings.

Figure 7.1 provides a tree-like (branching) representation of motives which can influence human attitudes towards the conservation and management of biota. The components of this tree are divided into effects that contribute to self-benefit and those that are purely altruistic. In practice, human assessments of policies for the conservation and management of biota frequently depend on a combination of the components (both self-interest and altruistic components) identified in Figure 7.1, as do assessments of the effectiveness of proposed policies. Sources of diversity in individual attitudes to the conservation and management of biota include variations in self-benefit expected by individuals, differences in

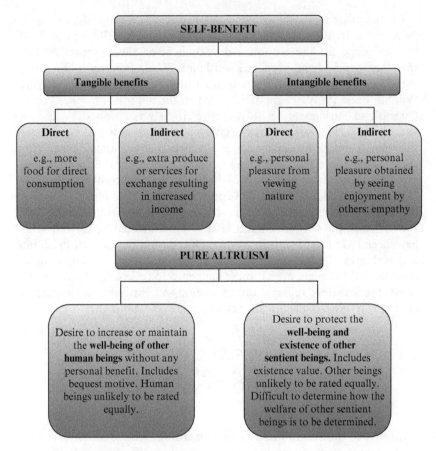

*Figure 7.1   Tree-like representation of factors that can influence
            the attitudes of individuals towards the conservation and
            management of biota*

their altruistic motives, and divergent views about the possible outcomes of such policies. Nevertheless, depending on the particular case, a high degree of social consensus can exist.

Altruism can reflect feelings of duty, such as a felt obligation to conserve wild species independently of one's own benefit from this action. However, this felt obligation need not extend to all species, and for most people, it does not. In Australia, for instance, the predominant view is that introduced species such as feral animals (pigs, goats, wild horses, deer, just to name a few) are pests and not worth protecting, whereas native species are. This is reflected in relevant legislation. One (but not the only) reason for this difference in attitudes is that the introduction of exotics tends to reduce local and global biodiversity. Nevertheless, as discussed later, some Australians do favour the conservation of selected introduced wild mammals, such as brumbies, in particular locations. Another bias is in favour of conserving species with humanoid attributes, such as mammals, particularly those that do not pose a direct danger to humans. There is considerable empirical evidence to support this hypothesis (see, for example, Tisdell, 2014, Chs 13 and 14, for evidence pertaining to the similarity principle and conservation preferences).

Different Values Resulting in the Same Objectives

As discussed in greater depth in Chapter 4 of this book, different values can lead to the same objective for managing biota. For example, the desire to conserve all species may be based upon the Kantian-like view that all species have a right to exist, and that human beings have a duty to facilitate this. Alternatively, the basis for pursuing this objective could be the view of Ciriacy-Wantrup (1968) that the potential utilitarian benefit from conserving all species at (at least) a minimum viable population substantially outweighs the cost of doing so. Both approaches, however, provide no guidance about what should be done if it is impossible to achieve their common objective (Tisdell, 1990), that is, they give no indication of which species or biota should have priority for survival. Furthermore, in the case of Ciriacy-Wantrup's approach, it does not seem that the cost of saving all species from extinction is low in relation to their possible utilitarian benefits to human beings. For example, the opportunity cost of conserving the orangutan seems to be very high in relation to potential utilitarian benefits of doing so, even if collective benefits are taken into account (Tisdell, 2015, Ch. 12). There is also the additional problem that the concept of a minimum viable population of a species is relativistic in nature (Hohl and Tisdell, 1993).

Pest Management

Attitudes to the conservation and management of species (and other forms of biota) that are sometimes regarded as pests raise several interesting issues. Their status may be assessed purely from the point of view of their benefits and costs to humankind or from a wider perspective. However, social conflict about the status of living things occurs when some individuals obtain a net economic benefit from their existence and others are harmed by this. Urban dwellers may obtain an estimated net economic benefit from an existing population of a wild species (for instance, elephants) but agriculturalists may suffer net economic losses from their presence (see, for example, Becker and Farja, 2016; Tisdell, 2015, Ch. 4). An anthropocentric economic approach to whether the population of a wild species should be conserved or reduced hinges on the net aggregate economic benefit of doing so, if the Kaldor–Hicks' (potential Paretian) approach to evaluating social gain is adopted (Tisdell, 2005, pp. 3–4). However, this approach does not take account of the distributional effect of the adopted management policy, for instance whether or not farmers should be compensated for damages caused by wildlife species if it is decided that it is socially desirable to conserve them.

When a species is designated as a pest some anthropocentric objections may be raised about this. First, doubts may be raised about the validity of the valuation. Secondly, it may be contended that the species has some beneficial effects in the ecosystem which have been overlooked. For example, it may be important in controlling another pest, and this benefit ought to be taken into account. Thirdly, the species might have positive future benefits to humanity along the lines suggested by Ciriacy-Wantrup.

Furthermore, ecocentric objections may be raised. For example, Aldo Leopold (1933; 1966) rejected the anthropocentric approach to wildlife management on the basis that all species have a role to play in 'the web of life' and believed that humankind should not destroy this web. This may be a reflection of a view that natural systems are a part of God's creation (or that of some supernatural force) and ought to be treated, if not with reverence, at least with respect independently of their benefits to human beings. The question then arises concerning whether this ecocentric point of view implies that there should be no intervention in natural ecosystems favouring or disadvantaging any species.

The answer to this question is probably not, if actual wildlife management policies are a guide in this regard. For example, where exotic species have been introduced to a country upsetting the previous balance of nature, it is not uncommon for the countries concerned to take action to limit the disequilibrating ecological impact of this development.

Animal Welfare Concerns

Animal welfare concerns are an additional factor complicating biological conservation and management. In some cases, they can limit the scope for conserving biodiversity. For example, animal rights activists may object to the use of the only available means to reduce populations of introduced feral mammals (in order to sustain wild biodiversity) on the grounds that they are inhumane.

Institutions and Environmental Conservation

The heterogeneity of nature and human behaviour in environmental conservation and management is also strongly influenced by institutional structures and developments, for example by the emergence of civil organizations representing different business or conservation interests, the nature of political structures and methods of economic organizations, such as the use of market systems as a means of economic management. The development of social institutions can result in different degrees of social embedding. They can follow and promote objectives which are only loosely related to those of their individual members and to some extent they are liable to become supra-organisms. Consider some of the influences of institutional arrangement on biological conservation. In this regard, attention will be given first to the influence of the market system and then to the impact of public organizations and NGOs on biological conservation.

7.4   MARKET SYSTEMS AND BIOLOGICAL CONSERVATION

Much attention has been given to ways in which market systems can result in economic failures in managing biota, for example because of environmental externalities, the public good/bad attributes of some biota, and open access to natural resources. While all of these factors can be significant, possibly of greater significance is social embedding caused by the development of market systems.

The view persuasively presented by Adam Smith in 1776 (see for example, the Everyman's edition of his Wealth of Nations, 1776 [1910]) is widely accepted, that market systems promote the common good provided that they are competitive. Moreover, it is argued that in such a system the pursuit of self-interest (usually interpreted as profit-maximization by producers and utility maximation by consumers) is socially desirable for achieving the common good, specified by Pareto (1927) as being a situation in which it is impossible to make anyone better off without making another worse off given the limited availability of resources. However, in some recent times, the justice of the system in terms of its income distributional consequences has been questioned by writers such as Rawls (1971) but defended by Richard Posner (1981) on the grounds that all tend to gain in the end because of the ability of the market system to promote economic growth (see Tisdell, 2009a, Section 4.1). Nevertheless, the general presumption today in most societies is that a market system is more desirable than its alternatives. While that seems to be so, this presumption provides a platform for prejudice and the pursuit of self-interest at the expense of the common good.

Once the idea is widely established that market systems are ideal means for managing resources (or societies), this becomes a type of social embedding because it is a commonly held principle. Those who benefit most from the system are likely to distribute propaganda and lobby to support it, even when they desire to restrict the competitive elements of it to their advantage. They are also liable to downplay the size and the significance of the failures of market systems, for example, in addressing demands for the conservation and management of biota if they profit from those failures being ignored. Hence, the development of market systems generates changes in generally held beliefs as do other changes in social organization. In addition, they create transaction cost barriers to altering the system. It becomes difficult for individuals or groups of individuals to alter the system when their change would be in the social interest. Both transaction costs factors and widely held beliefs become barriers to social change. Hence, a social lock-in effect occurs with the development of market systems.

The presumption that the market system is socially ideal has resulted in a reluctance of governments and communities to address market failures which adversely affect the conservation and management of biota. In short, it weakens the political will to do so. Consequently, allocative (microeconomic) inefficiencies (inefficient decisions about resource use and their conservation) are liable not to be addressed even when it is in the general interest to do so. Furthermore, as explained in Chapter 5 of this book, market systems create strong social pressures for economic growth (that is, for the production of marketed commodities) as a means of sustaining employment and increasing incomes. The levels of economic growth required to achieve these goals rise with population levels, the introduction of labour-saving technology (such as information technology) and growing income-aspiration levels. This economic growth is commonly favoured at the expense of the natural world. This poses a threat to the sustainability of human welfare.

It might also be noted that even when there are no obvious signs of market failure, the extension of markets (economic globalization) is a principal force in reducing global genetic diversity. This is brought about in several ways (see Tisdell, 2015, Chs 5 and 6). One of these is the result of increasing regional (international) specialization in production. Another is the increased divorce of human-created ecosystems (for example, agrosystems) from local environmental conditions. This also makes it possible to convert larger areas of natural ecosystem into human modified or altered ones, thereby increasing global biodiversity loss.

7.5   THE BEHAVIOUR OF PUBLIC ORGANIZATIONS AND NON-GOVERNMENT ORGANIZATIONS IN RELATION TO BIOLOGICAL CONSERVATION

Once established, public organizations and non-government bodies (as well as most social entities) often develop behaviours that do not exactly represent the wishes of all their members or, when they are the agents, the wishes of their principals. The situation is not only complicated by the bounded rationality of group decision-making but also by the fact that the principals of such bodies can have divergent demands. Moreover, it may also be difficult to identify the principals of NGOs. For example, the members of such organizations are not always clearly defined, and most of their donors may not be members of the organization. If the organization is large, the bulk of its supporters may exercise little or no control over the management of the NGO's affairs. To some extent the body becomes a supra-organism and its objectives may alter with the passage of time in a way at odds with the wishes of its founders and the bulk of its supporters.

Once established, most of these social entities seem to adopt their survival and frequently their growth as their prime objectives. This is an important influence on the trajectory of their actual behaviours. They are, however, a part of an interactive social system and this interdependence complicates and shapes their behaviours. For example, other social organizations may be developed or existing ones activated to oppose their goals. Consequently, the development and exercise of countervailing power becomes relevant (see Galbraith, 1973). Furthermore, social organizations, such as NGOs, face the possibility of growing competition for available resources from new NGOs or others that re-focus their mission to increase their 'market share'.

The existence of social organizations and their interdependence has a substantial influence on biological conservation and management. Just as individuals do not exist in a social vacuum, neither do social bodies and organizations. Social settings play a major role in determining the behaviour of public bureaucrats, civil society and business entities. The goals and behaviours of individuals are mediated by these settings. Consider these aspects further in relation to public bodies and environmental NGOs.

The Capture of Public Bodies, Public Administrators and Political Representatives by Special Interest Groups and Other Self-seeking Entities

Only a brief sketch can be provided here of ways in which non-government entities and individuals influence the existence and behaviour of public bodies and consequently their role in influencing environmental

conservation and management via their influence on politicians. Special
interest non-government entities often play a significant role:

- in determining the type of public bodies established and their
  mission;
- in bringing about changes in their mission;
- in ensuring senior officials appointed to those bodies are sympathetic to their goals or not (at least) hostile to these; and
- in obtaining the removal of influential personnel in public bodies
  who are perceived as posing a threat to the self-interest of these
  entities.

The ability of non-government entities to exert the above-mentioned
influences over the operation of public governance varies. The greatest
influence is likely to be exerted by those with large amounts of economic
resources and who are able to make the greatest net economic gain from
such intervention. These are generally entities that are able to benefit most
from the market system, including restrictions on the competitiveness of
the market system.

Apart from uneven direct influence of non-government entities on the
development and behaviour of the public sector, important indirect influences are also present. The behaviour of senior public officials is likely to
be influenced by the potential threat posed to their position or to their
public body by influential non-government pressure groups. Presumably in
their decision-making public officials take account of the relative political
power of such groups when their organizations have conflicting agendas.

Symbiosis may also exist between some public bodies and prominent
non-government entities. The latter may provide public support to sustaining a public body (and even expanding it) in return for the public body
championing its cause in government circles.

The dominance of market-related entities on government appears to
extend beyond state boundaries to include many international organizations such as the World Bank, the International Monetary Fund and the
World Trade Organization (WTO). The influence of national governments
on the policies of these UN bodies is very uneven, reflecting their own
perceived special interests, which in turn are liable to be a reflection of the
interest of dominant non-government entities (organizations) in the countries concerned. Understanding the reasons for policy proposals of such
bodies requires consideration of the social and political context in which
their policies are formulated.

On the whole, UN bodies have strongly advocated policies which
they believe will stimulate economic growth, and they have favoured

market-based solutions to economic problems. While they have given increasing attention to the possibility that economic growth can endanger the sustainability of economic growth because of its environmental consequences, they do not seem to favour substantially restricting economic growth because of this. Rather they have focused on policy measures designed to increase the compatibility of economic growth with environmental conservation, that is, so-called win–win policies. However, there appears to be no independent available overall assessment of how successful bodies of the UN have been in promoting environmental conservation goals globally. It is even possible that the policies of some of these bodies, such as the World Bank and the WTO, have had a net negative impact on environmental conservation. All of these bodies can be expected to propose policies that favour their continuing existence and potential growth.

Functions Performed by Environmental NGOs and the Diversity of their Goals

Environmental NGOs differ considerably in the type of functions they perform. Individual environmental NGOs may be involved in different combinations of the following activities (the first of which is mandatory):

- obtaining funds from government or private sources to fund their activities, and usually voluntary contribution of resources, such as labour and land, to support their agendas;
- advocacy, both political and legal;
- direct action to obstruct developments considered to be detrimental to environmental conservation (for example, Greenpeace);
- dissemination of information and education to create environmental awareness;
- engaging in environmental research or helping to fund relevant research of others;
- directly performing conservation activities, for example, by acquiring land for conservation purposes, organizing voluntary labour contributions for on-the-ground conservation activities; and
- providing advice, resources and assistance to landowners wishing to pursue environmental objectives.

Their missions also vary considerably. Some focus on the conservation of particular species (for example, the Australian Koala Foundation, the Yellow-eyed Penguin Trust in New Zealand) whereas others focus on the conservation of broader sets of biota (for example, WWF). They also differ in the areal extent of their involvement in activities supporting environmental conservation. The geographical boundaries of their activities can range from being extremely local, to being regional, national or global and display a huge variety of spatial patterns.

Possible 'Efficiency Failures' of Environmental NGOs

The 'efficiency failures' of environmental NGOs appear not to have been well researched. Therefore, my hypotheses in this regard need to be followed up by more empirical research. However, before outlining these suggestions, some discussion of what can constitute 'efficiency failure' in this context is needed.

Efficiency failures may be assessed from the point of view of the NGO itself or in a broader context. An assessment from a consequentialist point of view (the one most frequently adopted in economics) is to consider the relationship between the ends (mission) of an NGO, its set of available means, and the means actually used by it to achieve its ends and the subsequent results. From a consequentialist perspective, efficiency failures of entities (avoidable shortcomings in achieving their goals) are usually divided into two components:

- failure to minimize the costs of their activities (that is, cost or X-inefficiency); and
- allocative inefficiency, that is, failure to allocate their available resources optimally between their competing ends, resulting in their desires not being satisfied to the maximum extent possible.

These types of failure in the efficiency of decision-making are illustrated in Figure 7.2. For illustrative purposes, it is assumed that an environmental NGO wants to avoid a reduction in the population of two species and if possible increase their population. Let x_1 represent the entity's contribution to the level of the population of species one and let x_2 represent that for species two. Given its available resources, its production possibility function might be as indicated by the curve ABCD. If it has a well-defined preference function for contributing to the level of the population of the focal species (which satisfies transitivity conditions and assuming that the entity favours higher populations of both species), the indifference curves market I_1I_1, I_2I_2 and I_3I_3 provide a possible representation of its preferences. Efficiency in satisfying its wants requires it to act in a way which results in it being on the frontier ABCD at point C, the point at which there is optimal conformity between its trade-off possibilities in contributing to the population levels of the focal species and its comparative

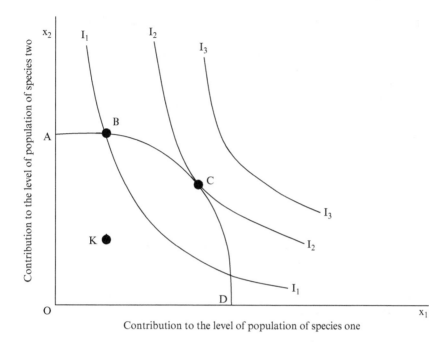

*Figure 7.2   An illustration of two types of efficiency failings which an
NGO may make in trying to conserve multiple species*

preference for doing so. Clearly, this would not be satisfied at point B, a
point corresponding to allocative inefficiency. It is possible, however, that
the activities of the entity may result in the entity operating at a point
below the frontier ABCD, for example, at point K. When this occurs,
X-inefficiency or cost-inefficiency is present.

An NGO may be at a point, such as K, because it allocates funds to
finding, treating and attempting to rehabilitate injured animals rather than
using its resources to obtain additional suitable habitat for the species,
although the latter strategy is more effective in maintaining or increasing
the population of the species concerned. Furthermore, a point such as B
may be reached by the NGO because its knowledge of its trade-off possibility function, ABCD, is poor.

The above discussion assumes that members of an NGO have a common
aim, that is, they act as a team. However, they may differ in their objectives.
Some, for example, may be more concerned with animal welfare (reducing
the suffering of injured wild animals) rather than increasing or preventing
a decline in their populations. Moreover, there is also the difficulty mentioned above that the membership of an NGO can be a fuzzy set, in which

case it is not clear whose preferences are to count and to what extent they should do so, or do so in practice.

It should also be noted that the type of rationalist consequential approach to decision-making outlined above is considered by some scholars to be too narrow for assessing human behaviour, particularly group behaviour. Reasons include its failure to take into account bounded rationality, social conflict within organizations, and the need to take account of dynamic factors such as flexibility to alter goals as circumstances change.

Some Possible Collective Efficiency Failures of Environmental NGOs

For reasons other than inefficiencies in their internal allocation of resources and cost-inefficiencies, environmental NGOs may collectively conserve a smaller number of species or types of biota (that is, less biodiversity) than is possible relative to the aggregate amount of resources they use. Possible reasons include:

- too many NGOs concentrating on conserving the same species (or set of species or biota) in order to maximize the donations or resources available to them individually; and
- competitive expenditure on advertising, image-building and other forms of promotion in order to seek donations (resources) which, after allowing for this expenditure, can result in a decline in the net amount of funds (resources) available to every NGO involved in this type of competition.

There appears to be a temptation for many NGOs to focus on conserving the most likeable of wild species, especially if potential donors can be convinced that they are in danger of extinction. The species involved tend to be human-like species, primarily mammals (Tisdell, 2014, Chs 13 and 14). Although both likeability and perceived danger of extinction influence the public's support for the conservation of wild species (Tisdell, 2014, Ch. 12), the presence of a high degree of likeability of one species (for example, the koala, *Phascolarctos cinereus*) may result in it obtaining by far the largest amount of funding for conservation compared to a likeable but much more endangered species (such as the hairy-nosed wombat, *Lasiorhinus kreffti*); even when a greater allocation in favour of the latter species seems to be socially desirable (Tisdell, 2014, Ch. 7).

Balkanization of conservation efforts by NGOs can also have other costs. For example, if economies of scale in administration are significant, balkanization will reduce the economic efficiency of their collective operations. In some instances also, there may be barriers to the entry of

NGOs to a field. Well-established and widely recognized NGOs may have an advantage in fund-raising compared to new competitive entrants.

It is also pertinent to note that public bodies (including international organizations) are usually active in promoting their self-interest in sustaining themselves and growing. A considerable proportion of their resources may be used for that purpose, for example, for image-building, exaggerating the value of their output and the demand for it. Consequently, the types of economic failures outlined above are not limited to NGOs.

7.6   CONSERVATION POLICIES AND ECONOMIC VALUATION TECHNIQUES: GENERAL COMMENTS

Economic techniques for the evaluation of the social desirability of environmental change have both strengths and shortcomings. It is not intended here to undertake a detailed assessment of the comparative virtues and shortcomings of particular types of these techniques. Rather it is intended to consider broadly the worth of revealed and stated preference techniques and their application to policy proposals. Nevertheless, at the outset, it should be noted that these are often the only available methods for measuring the strength of individual valuations of environmental change. They can therefore provide useful supplementary information about this, which otherwise would not be available. However, these estimates must be used with caution when formulating environmental policy. Further details on some of the topics covered briefly in this section are available in this book in Chapter 4, which concentrates on assessing techniques for economic valuation, and in Chapter 5, which pays particular attention to social embedding in contemporary societies.

The Validity of Economic Techniques Used to Obtain the Valuations by Individuals of Biological Changes

Reasons why care is needed in using (for assessment and policy purposes) economic valuations by individuals of biological changes include the following:

- They are subject to manipulation as a result of information provision. This may be deliberate or accidental. There is strong evidence that environmental valuations are significantly altered by information provision (Tisdell, 2014, Chs 6 and 11). Actual behaviours depend on beliefs, which in turn are influenced by the information individuals choose to gather or which they obtain independently of their actions.

- The Heisenberg-like uncertainty effect seems to be unavoidable when stated preference methods of eliciting preferences are used (Tisdell, 2014).

- When stated preference methods are used, individuals often have to construct their preferences or evaluations because their pre-existing preferences are incomplete. They usually have to do so in a limited time-period and with limited information. They may, therefore, be heavily influenced by information provided by the material used for elicitation purposes.

- Within relatively short periods of time, stated values are likely to be subject to drop-off or erosion effects (Tisdell, 2014).

- They are also subject to various forms of social embedding and the nature of this embedding can alter considerably during lengthy periods of time. Hence, these values are likely to be historically determined to a significant extent.

Shortcomings of Economic Techniques for Assessing the Social Desirability of Biological and Other Environmental Changes and their Limitations as a Guide to Public Policy

Most frequently, the monetary amount individuals are willing to pay or willing to accept as compensation for a biological change are summed to determine aggregate willingness to pay (or aggregate economic value) and these results are used in conjunction with the potential Paretian improvement criterion (also known as the Kaldor–Hicks criterion) to determine whether a biological change is socially desirable. Revealed preference approaches adopt similar aggregation methods. This type of summation procedure does not take account of the social consequences of environmental changes for alterations in the distribution of income; whether or not losers from such changes should be compensated for their losses; how this is to be done (if it is considered to be desirable); and how much compensation ought to be paid.

In addition, little attention is given to alternative methods of settling social conflict when some individuals gain and others lose as a result of environmental policies. The problem of maintaining social harmony (minimizing serious social disharmony) is virtually side-stepped, as are political constraints on the choice of environmental policies.

Furthermore, limited attention is given to the role of organizations in determining environmental behaviour and policy. The evolution of relatively impersonal forms of governance and societies consisting of a very large number of individuals (for example, states), has significantly limited the ability of individuals to influence collective policies. Individuals may have had greater influence on collective decisions in ancient times when social organization was based on decision-making by small bands of people, or tribes of a fairly small size. The main avenue open to individuals in contemporary societies for increasing (or having much effect at all on) collective decisions is by joining or supporting organizations which are able to represent their views and act as political pressure groups. Consequently, sociological aspects of group formation, their behaviours and their dynamics need to be studied in order to obtain a more complete picture of the determinants of environmental policy.

Knowledge Deficiencies and Logical Failures in Pursuing Conservation Goals, for Example, Failure to Account for Opportunity Costs or to do so Adequately

A problem of relying on valuations of biota by members of the general public for policy purposes is that the public are likely to be poorly informed about the role played by different types of biota in ecosystems and the nature and magnitude of the services (or drawbacks) attributable to them. The scientific community is likely to be better informed but not fully informed. Nevertheless, scientists may also be biased in their personal valuations. For example, they may strongly favour the survival of species or biota that are the prime focus of their research, and place much less emphasis on the conservation of other biota. Scientific policy advice is, therefore, unlikely to be free of personal values.

It is problematic how much weight ought to be placed upon the valuations of the general public of environmental changes, particularly when they are ill-informed. The populist path would be to just follow what the general public says it wants, if one demand is dominant. On the other hand, attempts might be made to better inform the public and then determine what its members want. However, there is a risk of the information provided being biased or deficient in other ways. There is no way to guarantee that this will not happen but open societies with freedom of speech can act as a useful antidote to this possibility.

Most members of the general public are not in a position to assess the opportunity costs and the more remote consequences of their decisions to support or oppose many environmental changes. Moreover, scientific evidence is often lacking on such effects and scientists and conservationists sometimes make errors in logic when giving policy advice about environmental conservation. Some scientists have suggested, for example, that those areas in which the density of a species is highest should be set aside for its conservation. For example Husson et al. (2008, p. 95) state that 'sites with the highest density of orangutans should be prioritized for conservation'. As, however, demonstrated in Tisdell (2015, Ch. 12), this may not minimize the cost of conserving orangutans because it fails to take account of the relative cost of conserving orangutans in different locations. The opportunity costs involved need to be explicitly taken into account. When that is done, it may happen that the most productive land for agriculture (or similar communal activities) should be set aside for conserving wildlife species because their densities are very high relative to the value of the land for agriculture, but this need not be the case.

Another dubious policy assertion is that the most productive land for agriculture be allocated to this use with less productive land being used for nature conservation (IUCN-UNEP-WWF, 1980; 1991). However, as shown in Tisdell (2005, pp. 34–8), this may result in a lower natural environmental output and less agriculture output than reserving better quality land for agricultural purposes. It depends upon the trade-offs or the opportunity costs involved. Nevertheless, it is pertinent to note that land which has the greatest commercial value will tend to be used for commercial purposes rather than for environmental conservation. This is because those who possess or can acquire this land are, as a rule, able to extract greater personal wealth from it by using it for commercial purposes rather than by conserving its natural biota. Therefore, if they are motivated only by their narrow economic self-interest, private landholders will choose to 'develop' their land even though conserving it constitutes its greatest social value.

Although it is important to take account of opportunity costs in assessing environmental changes, to do so can be quite complicated, especially if social rather than private opportunity costs are to be taken into account. This is because social opportunity costs can consist of subjective as well as objective elements. Moreover, the burden of social opportunity costs can be unequally distributed between individuals. Therefore, many of the problems which occur in social cost–benefit analysis do not disappear in cost-effectiveness analysis when the costs to be taken into account are social costs. Nevertheless, opportunity costs should not be ignored. One problem with many global valuations of natural ecosystems is that they give little or no attention to the opportunity cost of conserving such systems because they do not take into account the economic value of their transformation, partial or otherwise (see for example, Tisdell, 2015, Ch. 15).

7.7 CONSERVING KOALAS: CONFLICTS, VALUATION ISSUES AND ECONOMICS

Some specific examples will help to highlight several of the issues raised above. Most attention will be given to the protection and conservation of koalas but some other cases are also considered. The examples are mainly drawn from Australia but no doubt have their parallels elsewhere. Additional information about issues pertinent to the conservation of the koala is available in Tisdell et al. (2015).

The Status of the Koala

The koala is a charismatic, well-known species with humanoid features, which is liked very much by Australians and visitors to Australia. Its use in the Australian media, for example in children's books, has endowed it with considerable cultural significance. Although it is classified by the International Union for the Conservation of Nature (IUCN) as of least concern in Australia as a whole, it has in recent years been classified by the Australian Government as vulnerable in Queensland, New South Wales and the Australian Capital Territory. Its main stronghold is in Victoria. In some parts of its southern distribution it is regarded as a pest due to its adverse impacts on natural vegetation. For example, it was introduced to Kangaroo Island (located off the coast of South Australia) and has damaged stands of native tree species there. Attempts to reduce its population in areas where it creates ecological imbalance with native vegetation are not infrequently met with strong opposition from some members of the general public, given their empathy with this animal.

In contrast, in several local areas, particularly of Queensland and New South Wales, the koala is now extinct or on the verge of extinction. These are mostly localities where considerable urban expansion has occurred. The landscapes transformed by urban development have mostly been those containing prime koala habitat in which koala densities are very high. Not only does urban growth result in loss or degradation of this habitat but it increases the likelihood of koalas being struck by motor vehicles and being attacked by domestic dogs.

Parochialism in Conserving Koalas

Even though the continuing existence of koalas in Australia seems secure for the foreseeable future, several local NGOs and local government institutions have taken steps to protect it in their local areas. The geographical scale on which the koala is conserved is important to them. In other words, they want to conserve it locally even if it is secure nationally. Why might this be so?

There could be several reasons for this. For example:

- locals may not be convinced that the existence of the koala is secure nationally and may believe that their efforts will improve its chances of continuing existence;
- they may obtain utility from the local presence of the koala or knowing that it continues to exist locally;
- they may feel a moral obligation to act to conserve it and obtain moral satisfaction from doing so.

Despite efforts by local bodies, in many cases the koala is doomed to disappear locally. Why might the local bodies or organizations concerned about its conservation persist with their efforts to conserve it locally in this case? Possible reasons are that:

- they may not believe that their efforts will be to no avail;
- they may realize that their efforts will merely delay local extinction but may be morally compelled to act; and/or
- they may obtain extra utility from the presence of the koala in their locality for longer than otherwise.

An additional factor which ought to be taken into account is that once a local organization to conserve or protect koalas comes into existence, it can be expected to develop its own social dynamics. Social factors are likely to favour its persistence of organization even when it becomes clear that their 'original' mission cannot be accomplished. Therefore, this aspect also needs to be studied.

Excessive Funding of Koala Conservation Compared to that for Saving Critically Endangered Creatures

Both the likeability of species and their proneness to extinction are major influences on the public support for measures to conserve them (Tisdell, 2014, Ch. 12). Because the koala is greatly liked and well-known, a much larger amount of funds is available for its conservation than for many critically endangered species, such as the hairy-nosed northern wombat (Tisdell, 2014, Ch. 7). This bias is further reinforced by the existence of several NGOs having as their mission saving the koala and the absence of or limited presence of NGOs supporting the conservation of many highly endangered species.

Note that, given their focus on conserving particular species, many NGOs could be inclined to exaggerate the imminence of extinction of their focal species in order to increase public support for their campaigns. However, if this is overdone by an NGO, it risks losing credibility.

The Strategy of Rescuing, Treating and Rehabilitating as well as Translocating Koalas: An Inefficient Approach to Conservation?

Several policy interventions designed to protect and benefit koalas do not seem to be cost-efficient in contributing to their conservation. These include programmes for recovering sick or injured koalas from the wild, their hospitalization, potential treatment and return to the wild. Although relevant statistics are difficult to access, as reported in Tisdell et al. (2015), some are available from koala hospitals in south-east Queensland. These indicate that an average in excess of 1400 koalas per year were admitted to these hospitals between 1997 and 2010. The majority (around 1000 on average per year) were dead on arrival or euthanized. About 400 per year were successfully treated and returned to the wild.

The average cost per koala treated is said to be in the range of 380–1500 AUD (see Tisdell et al., 2015 for more information). If a low estimate of 400 AUD per koala treated is used, the total cost of this programme would have amounted to at least 2.24 AUD million for the period 1997–2010.

The survival rate of koalas after leaving hospital is unknown. They are, however, returned to the original place where they were found or as near as possible to this. Consequently, they are likely to face the same threats as they faced in the first place and, in all probability, their return to an environment in which the carrying capacity for koalas is already exceeded. These factors suggest that the programme is likely to do little to sustain koala populations already under stress.

One wonders whether from the point of view of conserving koala populations the sum of at least 2.24 AUD might have been better spent purchasing and protecting suitable areas of habitat for koalas. Nevertheless, in all probability strong public opposition would most likely arise if the hospitalization programme were abandoned in favour of this conservation policy. This is because considerable empathy has been established with the koala and individuals are emotionally stressed by the suffering of these injured animals. Consequently, conservation policies for the koala (and several other species) are complicated by human concerns about the welfare of individual animals. This can cloud their assessment of effectiveness programmes for the conservation of species and the welfare of their populations as a whole.

Historical Changes in Attitudes to Koala Conservation

The Australian public's support for the conservation of koalas seems to be much stronger now than in the past. In the 1920s, koalas were hunted commercially for their pelts (Moyal and Organ, 2008). Although some members of the public objected to this, it would be politically impossible to allow this to happen today, even if it could be shown to be commercially viable.

Public attitudes to the conservation of and the consumptive use of many species can alter substantially within a few decades. This has occurred in Australia, for example, in the case of marine turtles (see Tisdell and Wilson, 2012, Ch. 9) and also in the case of whales. Similar changes have also happened elsewhere in the Western world, for example, in the United States.

Many economic valuations of environmental phenomenon do not allow for time-related changes of the type mentioned above, although some economists (for example, Bishop, 1978) have stressed their importance for policy formulation. Uncertainty about future values is probably the main reason for this neglect. There is, however, a case for considering what the crystal ball might hold.

7.8   THE ROLE OF WILDLIFE REHABILITATION CENTRES IN NATURE CONSERVATION: OBSERVATIONS ABOUT THE CONSERVATION OF ASIAN ELEPHANTS AND ORANGUTANS

Rehabilitation centres for wildlife can *potentially* contribute to their conservation in at least three ways. These are:

- saving stricken animals, caring for them and successfully returning them to the wild to help maintain the populations of their species;
- when these centres are used for tourism, creating awareness among their visitors to such centres about threats to the survival of their focal species; and
- motivating visitors to take actions which assist the conservation of these species, for example, donating funds for conservation of the species, changing their own behaviours or political stance.

In many (but not all) cases, however, these centres do not appear to fulfil any of these roles effectively. Some act purely as a tourist attraction which contributes income to those involved with these centres. They can also add to incomes in the local economies where they are located (and to some extent beyond) via the economic multiplier effect and by the local sourcing of souvenirs often sold at such venues. A study conducted at Pinnawala Elephant Orphanage in Sri Lanka supports this proposition (Tisdell and Bandara, 2005). Visitors to this reserve learned little about the plight of the Asian elephant and were primarily entertained by the activities involving the rescued elephants (such as the bathing of the elephants, an elephant carrying a very large weight) (Tisdell and Bandara, 2009). These elephants had become domesticated. Moreover, there is no evidence of any of this centre's rescued elephants ever being returned to the wild. This is under-standable because those returned to the wild would be unlikely to survive.

In the wild, returned elephants may lack the skills to survive and might find it impossible to join any social grouping of elephants. Furthermore, they would be unlikely to add significantly to the surviving elephant population, which in many cases is already too high for the carrying capac-ity of areas where elephants are present. The same could well be so for orangutans released from rehabilitation centres; the numbers may not be large and the chances of their survival in the wild may be low, especially those rescued as babies who have not had time to learn from their mothers the skills of surviving in the wild.

This is not to suggest that wildlife rehabilitation centres have no benefits. They can have significant economic benefits: they do help stricken animals and satisfy humans who want to alleviate the suffering or misfortune of individual animals, frequently caused by the activities of other human beings. Their compassion may be a form of atonement for the misdeeds and shortcomings of others.

7.9   ECOTOURISM ENTERPRISES AND THE CONSERVATION OF SPECIES

At least two criteria have been stressed in relation to assessment of eco-tourism undertakings, namely:

- their ability to contribute to the conservation of biota that are the focus of their endeavours; and
- their potential to generate an economic benefit, especially for local communities.

There is no doubt that ecotourism can have substantial local economic benefits, as examples in Tisdell and Wilson (2012) demonstrate. However, the *direct* contribution of ecotourism bodies to the conservation of their focal species is frequently negligible because the size of the population of species saved by these enterprises is very small compared to numbers needed to achieve a minimum viable population. For example, the number of albatross saved by the protected rookery of the Otago Peninsula Trust in New Zealand is insufficient to conserve the particular species targeted. Nevertheless, it makes a significant economic contribution to the local economy by attracting tourists to its facility (see Tisdell and Wilson, 2012, Ch. 13). Its indirect effect on generating support for the conservation of albatross is unknown. However, the type of tourism involved has no apparent negative effects on the conservation of albatross. Therefore, its benefits are clearly positive.

7.10　CONFLICTS BETWEEN CONSERVATIONISTS ABOUT PRIORITIES FOR THE SURVIVAL OF SPECIES: THE CONSERVATION OF FERAL HORSES (BRUMBIES) IN THE AUSTRALIAN HIGH COUNTRY VERSUS PRESERVATION OF ITS NATURAL ENVIRONMENT

An interesting recent complication (reported in interviews by Kerry Straight, 2016) in Australia has arisen between organizations that wish to conserve the population of brumbies in Kosciusko National Park and those who want to reduce their populations substantially (or even eliminate them in this area) because of their perceived threat to some rare endemic species in this alpine habitat and their adverse impacts on soil erosion. Those in favour of their conservation want them conserved at a population level higher than that proposed by the New South Wales Government. They do so mainly on the basis that these wild horses are of significant cultural value for Australians and seeing them is welcomed by some tourists. Some lobbyists wanting to save brumbies consider that shooting them is inhumane and object to culling on this basis also.

　The extent to which economic valuation methods could be used to resolve this conflict is unclear. After informing respondents of the environmental damages caused by these wild horses, one might in principle determine the willingness of Australians to pay for the conservation of different levels of populations of brumbies in Kosciusko National Park. However, the information provided about the amount of environmental damage caused is liable to be subject to dispute. Furthermore, 'buying off' those who oppose the plan to greatly reduce the population of brumbies is unlikely to be morally acceptable and politically not workable. It is doubtful whether the Kaldor–Hicks approach (with compensation being paid to the aggrieved parties) could be a practical solution for resolving this type of conflict.

There are many different species of feral animals in Australia. They have primarily been regarded in Australia as pests and are not afforded state protection, unlike most endemic species of wildlife. However, it is possible that if some feral species were now driven to the brink of extinction there would be public support for their conservation. For instance, this could be so in the case of the dromedary camel (*Camelus dromedarius*), which played a significant role in the early development of inland Australia, and which has attained some cultural and historical significance.

7.11   CONCLUSIONS

Biological conservation and management is multi-faceted and influenced by diverse motives and perceptions. This makes the economic valuation of environmental change very challenging. While market systems enable large economic systems to be managed systematically, the idea that they result in a social optimum is far-fetched. Apart from the presence of standard types of market failure (which are well covered in the available literature), they are a source of significant social embedding in contemporary societies. This can reinforce the economic failures of market systems, for example, in promoting environmental conservation.

It is difficult in large modern societies (where most individuals know only a very small fraction of members of these societies) for individuals to have any significant influence on the supply of public goods or those requiring collective action except through organizations which may represent their views. Therefore, it is vital to study the role of organizations in influencing economic and environmental change. It has been argued that such organizations are subject to biases and economic failures. For example, public bodies, politicians and public servants are subject to capture by special interest groups and mostly these groups are those who benefit economically from the market systems or interference with it. International public bodies are also unlikely to be free of such influences and social embedding.

NGOs, particularly in Western countries, are very active in promoting environmental conservation goals. However, as pointed out, their competition to obtain funds and resources is liable to result in cost and allocative inefficiencies, as well as collective failures in organizing funds and resources for conservation purposes. For example, too many may concentrate on conservation initiatives for which funds are most forthcoming (Tisdell, 2009b). This is similar to the phenomenon observed when there is open access to natural resources.

Although methods of economic valuation can provide useful information for assessing environmental change, they also suffer from several limitations. Possibly their most important failure is their lack of guidance about how best to solve political problems involving social conflict.

The specific examples considered in this chapter (the conservation of koalas, Asian elephants and orangutans, albatross and brumbies versus conservation of the natural environment and native biota in the alpine area of Australia) illustrate the complexities of valuing environmental changes and resolving social conflict. On the one hand, these examples show how economic analysis can pinpoint the nature of cost and allocative economic inefficiencies which can occur in programmes for the conservation of species, for example the koala. On the other hand, they highlight complications (seemingly not amenable to economic analysis) arising from mixed motives in protecting wild animals, for example, the demand to provide animals from socially favoured species with help and humane treatment if injured or stricken in some way, irrespective of the effectiveness of this in conserving the species concerned. It seems that many conservation policies are not based on consequentialism but are emotionally based. Furthermore, the examples given in this chapter highlight the importance of taking into account the role of organizations when considering policy issues involving environmental changes. This role has largely been neglected by economists.

REFERENCES

Becker, N. and Y. Farja (2016), 'The cattle–wolf dilemma: interactions among three protected species', *Environmental Management*, 1–14.

Bishop, R.C. (1978), 'Endangered species and uncertainty: the economics of a safe minimum standard', *American Journal of Agricultural Economics*, **60**(1), 10–18.

Ciriacy-Wantrup, S.V. (1968), *Resource Conservation: Economics and Policies*, 3rd edn, Berkeley, CA: Division of Agricultural Science, University of California.

Galbraith, J.K. (1973), *Economics and Public Purpose*, Boston, MA: Houghton Mifflin.

Hohl, A. and C.A. Tisdell (1993), 'How useful are environmental safety standards in economics? The example of safe minimum standards for protection of species', *Biodiversity and Conservation*, **2**(2), 168–81, reprinted in C.A. Tisdell (2002), *The Economics of Conserving Wildlife and Natural Areas*, Cheltenham, UK and Northampton, MA, USA: Edward Elgar Publishing.

Husson, S.J., S.A. Wick, A.J. Marshall, R.A. Dennis, M. Acrenaz, R. Brassey, M. Gumel et al. (2008), 'Orangutan distribution, density, abundance and impacts of disturbance', in S.A. Wick, S.S.U. Atmuko, T.M. Satia and C.P. Scheik (eds), *Orangutans: Geographical Variation in Behavioural Ecology and Conservation*, Oxford: Oxford University Press, pp. 77–96.

IUCN-UNEP-WWF (1980), *The World Conservation Strategy: Living Resource

Conservation for Sustainable Development, Gland: International Union for the Conservation of Nature and Natural Resources.
IUCN-UNEP-WWF (1991), *Caring for the World: A Strategy for Sustainable Living*, Gland: World Conservation Union.
Kumar, P. (ed.) (2010), *The Economics of Ecosystems and Biodiversity: Ecological and Economic Foundations*, London, UK and Washington, DC, USA: Earthscan.
Leopold, A. (1933), *Game Management*, New York: Scribner.
Leopold, A. (1966), *A Sand County Almanac: With Other Essays on Conservation from Round River*, New York: Oxford University Press.
Millennium Ecosystem Assessment (2005), *Ecosystems and Human Well-being: A Synthesis*, Washington, DC: World Resources Institute.
Moyal, A. and M. Organ (2008), *Koala: A Historical Biography*, Collingwood, Victoria: CSIRO Publishers.
Pareto, V. (1927), *Manuel d'Economie Politique*, 2nd edn, Paris: Giard.
Posner, R.A. (1981), *The Economics of Justice*, Cambridge, MA and London: Harvard University Press.
Rawls, J.R. (1971), *A Theory of Justice*, Cambridge, MA: Harvard University Press.
Smith, A. (1776 [1910]), *The Wealth of Nations*, Everyman's edn (first published 1776), London: Dent and Sons.
Straight, K. (2016), 'Brumbies', *Landline*, ABC, accessed 13 September 2016 at http://www.abc.net.au/landline/content/2016/s4514396.htm.
Tisdell, C.A. (1990), 'Economics and the debate about preservation of species, crop varieties and genetic diversity', *Ecological Economics*, **2**(1), 77–90, reprinted in C.A. Tisdell (2002), *The Economics of Conserving Wildlife and Natural Areas*, Cheltenham, UK and Northampton, MA, USA: Edward Elgar Publishing.
Tisdell, C.A. (2005), *Economics of Environmental Conservation*, 2nd edn, Cheltenham, UK and Northampton, MA, USA: Edward Elgar Publishing.
Tisdell, C.A. (2009a), *Resource and Environmental Economics: Modern Issues and Applications*, Singapore, Hackensack, NJ, London and Beijing: World Scientific.
Tisdell, C.A. (2009b), 'Institutional economics and the behaviour of conservation organizations: implications for biodiversity conservation', in K. Ninan (ed.), *Conserving and Valuing Ecosystem Services and Biodiversity: Economic Institutional and Social Challenges*, London and Sterling, VA: Earthscan, pp. 175–91.
Tisdell, C.A. (2014), *Human Values and Biodiversity Conservation: The Survival of Wild Species*, Cheltenham, UK and Northampton, MA, USA: Edward Elgar Publishing.
Tisdell, C.A. (2015), *Sustaining Biodiversity and Ecosystem Functions: Economic Issues*, Cheltenham, UK and Northampton, MA, USA: Edward Elgar Publishing.
Tisdell, C.A. and R. Bandara (2005), 'Wildlife-based recreation and local economic development: the case of Pinnawala Elephant Orphanage in Sri Lanka', *Sri Lanka Journal of Social Sciences*, **28**(1&2), 65–96.
Tisdell, C.A. and R. Bandara (2009), 'A Sri Lankan elephant orphanage: does it increase willingness to conserve elephants? How do visitors react to it?', in J.D. Harris and P.L. Brown (eds), *Wildlife Destruction, Conservation and Biodiversity*, New York: Nova Science Publishers Inc., pp. 253–76.
Tisdell, C.A. and C. Wilson (2012), *Nature-based Tourism and Conservation: New*

Economic Insights and Case Studies, Cheltenham, UK and Northampton, MA, USA: Edward Elgar Publishing.

Tisdell, C.A., H. Preece, S. Abdullah and H. Beyer (2015), 'Parochial conservation practices and the decline of the koala: a draft', *Economics, Ecology and the Environment*, Working Paper No. 200, Brisbane: School of Economics, The University of Queensland.

8. Climate change: general aspects, and alterations in energy sources and use as responses

8.1 INTRODUCTION

Effectively addressing the numerous policy issues raised by global warming and climate change is a major challenge of this century (Ross et al., 2016). Many of the issues involved are difficult to resolve politically because of conflicts between individual and collective self-interest, as exemplified by the well-known prisoner's dilemma problem discussed in the theory of games. This problem also has an international analogue because it is usually not in the selfish interest of individual nations to adopt measures to restrict their GHG emissions (the prime cause of contemporary global warming) in order to reduce these emissions globally. Similarly, if guided only by their own self-interest rather than collective global benefits, individual nations have no incentive to maintain or increase their sequestration of carbon dioxide (CO_2).

This problem can be compounded by international trade. For example, nations exporting fossil fuels have an economic incentive to export these even though they will add to GHG emissions in the importing countries. Likewise, importing countries continue to import materials (such as palm oil, timber and several other products) knowing that this will reduce carbon sequestration (due for example, to reduced forest cover) in nations exporting these commodities. In such cases, neither exporters nor importers may feel morally responsible for the environmental consequences of their trading decisions. Divided responsibility for adverse environmental and social behaviour often results in no one feeling morally responsible for it (Tisdell, 1989; 1990).

Only a few of the many socio-economic issues raised by human-induced global warming and climate change can be considered in this chapter. Attention will be given in turn to the following topics:

- Failure to control 'adequately' CO_2 emissions due to social embedding and lack of foresight intelligence.

- Major biophysical processes resulting in climate change and reasons for concern about the use of fossil fuels and other sources of CO_2 emissions following the start of the Industrial Revolution.
- The prospects generally of reducing CO_2 emissions by substituting energy sources by energy savings, and by greater sequestration of CO_2.
- The scope for substituting the types of resources used for electricity production in order to reduce CO_2 emissions, and the economics of this substitution.
- Economic strategies to lower fossil fuel use in order to reduce GHG emissions.
- Additional sustainability issues (for example, aspects of the availability of stocks of energy resources and their depletion) and the importance of the holistic environmental valuation of alternative forms of energy supply.

8.2   FAILURE TO CONTROL 'ADEQUATELY' CO_2 EMISSIONS DUE TO SOCIAL EMBEDDING AND LACK OF FORESIGHT INTELLIGENCE

Despite mounting evidence about the adverse environmental consequences of global CO_2 emissions caused by the nature of human activity (mostly economic activity), the problems involved have not been effectively addressed by governments and by most individuals. To a large extent, modern economies are economically and technologically locked into practices which foster continuing CO_2 emissions and its accumulation in the atmosphere. This is resulting in:

- global warming;
- climate changes;
- rising sea levels; and
- increased acidification of water bodies.

These environmental changes have already started and are expected to become more pronounced in a relatively short period of time. They will threaten global economic sustainability, the biological diversity of organisms and have other adverse effects, for example, on human health.

Ross et al. (2016, p. 363) attribute the failure to 'adequately' limit CO_2 emissions to lack of foresight intelligence and state that 'when it comes to confronting environmental perils that lie in the future and unfold gradually, our species has generally failed to exercise *foresight intelligence* – that is, to recognize, diagnose, plan and not to address these perils before it is too late to do so'.

The Kyoto Protocol failed to prevent rises in anthropocentric global CO_2 emissions and did not halt reductions in the size of biological sinks for sequestering CO_2. It is, however, possible that it moderated changes in these variables. It remains to be seen whether the Paris Agreement will be more successful in this regard.

The evolution of large political units and the increasing prevalence of the market system (as a result of economic development) may have made it more difficult than in earlier times to exercise foresight intelligence and may have increased social embedding thereby making it increasingly difficult to respond adequately to current environmental challenges, particularly global warming concerns and GHG emissions. Economic growth has fostered the rise of nation states. Market systems play a major role in the organization of resource use by nation states and in their survival. Furthermore, market systems have been an effective means for stimulating economic growth. However, the nature and magnitude of this economic growth seems to be environmentally and socially unsustainable. For example, given the current forms of social embedding, an eventual environmental crisis due to GHG emissions may be unavoidable. This will also result in an economic crisis. Therefore, it is necessary to consider features of social embedding in modern societies which could lead to this result.

Even though CO_2 emissions are not the only types of gases (released by human activity) contributing to global warming, they are the major contributor. In any case, the social embedding problems mentioned here apply to failure to 'adequately' control the release of all GHGs. The two main mechanisms driving social embedding or lock-in of human behaviours are:

- the development of shared values by members of societies; and
- the evolution of organizational or institutional structures which constrain collective decision-making (Tisdell and Svizzero, 2017).

These elements are usually not independent. For example, major beneficiaries from the market system usually promote the idea that the pursuit of personal or corporate self-interest is in the social interest. The widely held view in contemporary societies that liberalism is socially desirable seems to have developed hand-in-hand with the evolution of market systems. Thus, personal economic gain is seen as a desirable social value and a sign of success. This is also reinforced by the idea that economic growth is a valuable social goal. These values contribute to contemporary social embedding (see Chapter 5 of this book).

Structural Lock-in and CO_2 Emissions

Structural lock-in can take a variety of forms. At this stage of economic development these forms make it very difficult to reduce CO_2 emissions substantially. *Technological path dependence* is one such problem. Many modern technologies have been specifically developed to rely on fossil fuels, such as internal combustion engines. An across-the-board major reduction in the use of fuels generating CO_2 emissions in a short period of time would probably result in negative economic growth and a major decline in incomes and human welfare.

There are also relevant types of social structural lock-in which limit our collective ability to control CO_2 emissions. For example, there is *the prisoner's dilemma type of problem*. While it may be in the collective interest of all nations and most individuals to limit CO_2 emissions, each has an individual incentive not to honour any agreement to do so. In this case, if others keep to the agreement, each party will be better off by not honouring the agreement. The same is the case if others do not honour the agreement. The result is that the agreement fails and all are worse off than they could be by honouring the agreement. The problem at present is that no measures are in place for penalizing nations that fail to keep to an agreement and reduce their CO_2 emissions. Furthermore, each nation maintains its own sovereignty and does not allow independent monitoring of its emissions.

Concern for Future Generations

Since the environmental effects of human elevation of CO_2 levels are long-lasting and many of its adverse consequences are delayed, its impact on future generations will be greater than on current generations. Therefore, decisions by current generations to limit these emissions should logically depend on their concern for the welfare of future generations. In modern societies, such concerns may not extend beyond one's children and grandchildren and possibly great-grandchildren. Pearce (1998) has suggested that concern for the welfare of future generations be measured by a coefficient of concern but little has been done to specify empirically such coefficients. However, accurate measurement of this phenomenon is likely to be difficult. Tisdell (2015, pp. 68–9) examines various aspects of this issue, which have also been considered by Schelling (1995).

Ross et al. (2016, p. 364) suggest that this coefficient is likely to be low in modern societies, but it is unclear from their text whether it is lower now than in early tribal societies. This is because they argue that many tribes (for example, some Native American tribes) had a tradition of thinking multiple generations ahead but, on the other hand, the authors claim (p. 364) that this coefficient was low because 'our species evolved as a small group of animals with a focus on our own survival and short-term needs, along with those of our offspring and near kin'. This latter idea is similar to that of Hobbes (1651 [2010]) that early humans lived like animals. It may, however, be quite misleading to generalize in this respect (Svizzero and Tisdell, 2016) and hard to provide adequate evidence in support of this hypothesis. Nevertheless, Ross et al. (2016) express the opinion that the early animal-like existence of *Homo sapiens* has conditioned humanity to have a low and very selective coefficient of concern for future generations. This concern is usually limited to close kin.

It should also be noted that the concern of current generations for the welfare of future generations will be influenced (as far as CO_2 emissions are concerned) by their predictions about the future effects of CO_2 emissions and the ability of future generations to cope with these. Rational decisions therefore depend not only on the values of individuals but also on their view about how reality will unfold (Tisdell, 1968, Ch. 2). The scope for errors about the latter aspect depends on how well individuals are informed. However, even the well informed can be expected to be uncertain about future events involving CO_2 emissions and the development of new technologies. That is not to say they are completely ignorant but to emphasize that bounded rationality is a serious problem in precisely determining the long-term trajectories of the consequences of CO_2 emissions, as will be apparent from the last chapter in this book.

The Noisy-signal Problem, Wishful Thinking, and the Large Numbers Problem

Ross et al. (2016) suggest that changes in current weather patterns do not convince a host of individuals that global warming is occurring because these changes are erratic. Wishful thinking and denial of climate change is another problem raised by Ross et al. (2016). They point out that climate change denial is given greater credence when some scientists, politicians and special interest groups pander to this predisposition.

Ross et al. (2016) label this problem as the 'drop-in-the-bucket' problem. Reducing GHG emissions requires billions of people to cooperate in bringing it about. This cannot be effectively achieved by individual effort because the effect of each individual is miniscule and each is likely to regard their individual actions as ineffective from a pragmatic point of view. Although some individuals may act in this way as a moral imperative, it is highly unlikely that sufficient individuals and the sum of their actions will be enough to significantly reduce GHG emissions. This is why

collectively organized effort is required but there is no guarantee (as indicated above) that this will be forthcoming. Ross et al. (2016) outline various
policy measures that might be taken to overcome the above-mentioned
socio-economic barriers to reducing GHG emissions, even though none
guarantee success.

8.3   BIOPHYSICAL PROCESSES AND REASONS FOR CONCERN ABOUT THE USE OF FOSSIL FUELS

In order to appreciate the challenges posed by CO_2 additions to the
biosphere caused by economic developments since the beginning of the
Industrial Revolution (and which are still continuing), consider some of
the basic biophysical relationships involved and their evolution, bearing
in mind that the nature of some of these relationships are imperfectly
known and complex. In addition, changes in the accumulation of GHGs
in the atmosphere (of which CO_2 is the one of major concern) are not
the only influences on climate change during long periods of time. Their
effects on climate are superimposed on climate alterations which can
arise for other sources as specified below. In this section, biophysical
aspects of CO_2 accumulation in the atmosphere following the Industrial
Revolution will be discussed. This is followed by a brief discussion of
biological and geological carbon cycles, some evolutionary aspects of the
former and the disruption of those cycles by economic growth.

Biophysical Aspects of CO_2 Accumulation in the Atmosphere Following the Industrial Revolution

CO_2 levels in the atmosphere have risen from about 280 parts per million
(ppm) at the beginning of the Industrial Revolution to 401 ppm today
(September 2016) at the Mauna Loa Observatory, Hawaii (see Earth's CO_2
Home Page, 2016) and this concentration is still rising. Increasing CO_2
concentrations in the atmosphere (other biophysical factors held constant)
result in global warming and climate change. In addition, they increase
the acidity of water bodies, particularly the oceans (see Chapter 10 of this
book). Increasing atmospheric concentration of CO_2 (together with that
of some other gases) reduces the radiation of heat from the earth, causing
the temperature of the atmosphere to rise. A fall in atmospheric CO_2
concentration has the opposite consequence. Rising global temperatures
(apart from causing climate change) trigger sea level rises by melting sea
ice and glaciers, and to a lesser extent by the expansion of sea water. The

latter effect is a result of the warming of oceans due to heat transfer from the heightened temperature of the atmosphere.

Reasons why CO_2 levels have increased in the atmosphere following the start of the Industrial Revolution include the following:

- Greater use of fossil fuels which previously stored CO_2 below the earth's surface.
- A reduction in the biomass of flora which previously sequestered carbon.
- The accelerated loss of the carbon in the soil as a result of speedier decomposition of organic matter due to the extension of agriculture, and greater intensity (frequency) of cultivation of cropped land. The warming of the soil has also contributed to this effect.
- Greater use of calcite to manufacture commodities, for example, cement.

There have been major alterations in global and regional temperatures and climate in the past. Long periods of significant global warming and of cooling have occurred. While natural changes in concentrations of CO_2 in the atmosphere have sometimes contributed to these variations, other influences have also been important. These include long-term periodic changes in the position of the Earth in relation to the sun result-ing in Milankovitch cycles of global cooling and warming (Stacey and Hodgkinson, 2013). Particles from major volcanic eruptions have also been credited with causing regional and global cooling by blocking out incoming heat from the sun.

Starting about 12000 years ago, the Earth transited from a period of glaciation to one of warmer temperatures. This marked the beginning of the Holocene era. Although global temperatures have on average been ele-vated during this period, significant variations have also occurred, as well as substantial changes in regional climate during this era (Cunliffe, 2015).

From about 10000 years ago (the start of the Anthropocene era) onwards, agriculture began to develop in different parts of the world (Svizzero and Tisdell, 2014). This was a major event in the relationship between human-kind and the environment. The activities of some hunter-gatherers had significant environmental impacts; for instance, their activities extirpated several wild species and in some cases modified ecosystems. For example Australian Aborigines altered landscapes by the controlled use of fire to favour some wild species, thereby attracting them to areas of sprouting grass following fire, in order to make it easier to hunt them. However, these environmental changes were relatively minor compared to those that even-tuated as a result of agricultural development. Following the Agricultural

Revolution, the level of global population and economic activity trended upwards and accelerated after the start of the Industrial Revolution – a revolution that would have been impossible in the absence of the earlier Agricultural Revolution.

As a result of this economic growth and population explosion, humans have had a major impact on the global environment. They are, among other things, altering the globe's climate. Ironically, it may be environmental changes due to rapid climate change rather than resource depletion which will stymie economic growth and undermine the Earth's ability to support (at a satisfactory level of income) the current high level of the world's population, which still continues to increase. Sea level rises will reduce the land mass available for agriculture and much alluvial land will be inundated and subject to salting as a result of climate change. The extent to which agriculture will be able to shift to new regions and maintain or increase production to compensate for adverse climate conditions in other regions is uncertain (see Chapter 9 of this book).

Natural ecosystems are already being disrupted by global warming and climate change, and ocean acidification is bound to affect these systems further. Wild biodiversity loss can be expected to mount given the rapidity of contemporary climate change. Evolutionary biological adjustments will be unable to keep pace with the speed of these environmental developments. Furthermore, on land, there are major barriers (as a result of economic development) to the migration of many wild species in response to climate change. Most natural and protected areas have effectively become isolated islands which imprison several species within them.

Furthermore, no suitable climatic zones may continue to exist for the survival of some species. For example, those species are likely to disappear which are obligated to arctic conditions for their survival (such as polar bears and some species of penguins), or to high latitude conditions involving ice and snow (see previous chapter). Conceivably, some heritage varieties of crops will also disappear because no climatic areas remain where they can be grown successfully (see Chapter 9).

Biological and Geological Carbon Cycles

Biological and geological processes govern the concentration of CO_2 in the atmosphere. These processes have been disrupted by economic activity following the Industrial Revolution. Economic changes have resulted in reduced amounts of carbon being sequestered in biota and in geological deposits, thereby increasing the quantity circulating in the atmosphere, some of which has been absorbed by the oceans.

The evolution of biota performing photosynthesis (for example, trees and other plants, various unicellular species) has been responsible for reducing the concentration of CO_2 in the atmosphere and for increasing the atmospheric level of oxygen. CO_2 is stored as carbon in the biomass of such organisms. After biota die, their decomposition adds CO_2 to the atmosphere. When several forms of biological mass are reduced (for example, by burning to make way for agriculture), a large amount of carbon dioxide previously stored in this mass is added to the atmosphere.

Over several millennia, carbon has been removed from the carbon cycle by the deposition of organic material in the earth where it has formed fossil fuels, such as coal and oil. Since the Industrial Revolution humans have made increasing use of these deposits to provide new sources of energy, thereby adding large amounts of CO_2 to the atmosphere. Not all of these additions can be absorbed by the oceans (although they are a major sink) or by plants and other biota performing photosynthesis. This is so even though some autotrophic biota increase their mass in the presence of elevated CO_2 levels, thereby somewhat increasing their storage of carbon. Consequently, atmospheric CO_2 levels are rising and are accompanied by global warming and climate change.

8.4 COMBATING CLIMATE CHANGE BY SUBSTITUTING ENERGY SOURCES, ENERGY SAVING, AND GREATER BIOLOGICAL SEQUESTRATION OF CO_2

There can be little doubt that increased dependence on fossil fuels (first coal, then oil and subsequently natural gas) during and after the Industrial Revolution enabled the globe to support a massive increase in human population at a higher standard of living than otherwise would have been possible by depending only on earlier sources of energy, for instance, fuel supplied primarily by wood. The supply of biological renewable resources used for the production of energy at the beginning of the Industrial Revolution was quite limited. The use of fossil fuels overcame this constraint and reduced the significant deforestation which occurred during the commencement of the Industrial Revolution because of the high demand for wood to generate steam power. Fossil fuel use did, however, result in increased local and regional air pollution (for example, high levels of particulates in the air from burning coal). While these externalities have been moderated by technological developments, GHG problems arising from the use of fossil fuels have not yet been fixed technologically.

Although the use of fossil fuels has improved living standards and has supported a massive increase in global population, the continued use of fossil fuels poses a threat to the long-term sustainability of the current standard of living of most human beings and the ability of the Earth to sustain projected increases in human population. Fossil fuel use is creating an emerging and worrying carrying capacity problem for humankind. Therefore, urgent consideration is needed of the prospects for supplying global energy requirements by means other than by using fossil fuels and/ or by achieving a reduction in energy demand. Moreover, the limited scope for greater sequestration of CO_2 needs to be taken into account.

Basically, three different strategies exist for reducing increases in CO_2 in the atmosphere as a result of human activity. These are:

- substituting sources of energy with a high intensity of CO_2 emissions by those with a lower level of CO_2 emissions;
- reducing the demand for the use of energy; and
- increasing the sequestration of CO_2.

Alternative means of producing electricity other than by the use of hydrocarbons (for example, by using solar or wind power) have become more profitable due to technological progress. Taking into account the whole life-cycle of carbon use by these alternative systems, their adoption is socially more attractive than the use of coal-based generation of electricity in particular, and even more so when all of the adverse environmental externalities from the use of coal are taken into account. Market and political forces are likely to result in increasing substitution of alternative and less polluting sources of energy for fossil fuels, especially for electricity production. Nevertheless, as discussed below, challenging social and political problems are involved in this transition.

The rate of growth in demand for energy use can theoretically be reduced in several different ways. These include reducing population growth, lowering the amount of consumption of commodities requiring significant amounts of power, and by technological progress that decreases the amount of energy needed to supply goods and services. In recent times, the energy requirements of many household appliances and motor vehicles have fallen. However, because the number of these items worldwide is growing, total household energy consumption continues to rise.

While increased sequestration of CO_2 in organic matter is possible, the scope for storing CO_2 in this way appears to be limited. Policies based on this strategy are unlikely to be very effective on their own as a way of offsetting completely current and predicted levels of CO_2 emissions. They can, however, usefully supplement the other two strategies. These three strategies will now be assessed, paying particular attention to the supply of electricity.

8.5 CHANGING THE TYPE OF RESOURCES USED FOR ELECTRICITY GENERATION IN ORDER TO REDUCE CO_2 EMISSIONS

Several alternative means of producing energy rather than by the use of fossil fuels (coal, oil and natural gas) are currently available and have been adopted to varying degrees. These include the use of:

- solar power;
- wind power;
- hydropower;
- geothermal;
- tidal;
- nuclear;
- biofuels; and
- biomass.[1]

Partly as a result of efforts to produce 'cleaner' energy and to increase reliance on more sustainable sources of energy, the global use of these alternatives has risen in modern times, albeit at different rates. The rate of growth in the employment of solar and wind to produce electricity has been especially marked and has become increasingly competitive with the supply of electricity based on hydrocarbons as inputs. This is creating concerns for owners of coal-fired power stations who appear to have increased their lobbying against the use of wind and solar. At the same time, several generators of electricity have started to diversify their sources of production by including sustainable sources of energy supply in their electricity mix, as have some electricity distributors.

Comparative GHG Emissions from Producing Electricity by Alternative Means Taking into Account Complete Life-cycles of Supply

When the whole life-cycle and chain of electric power supply is assessed, all the possible sources of supply contribute, in varying degrees, to GHG emissions. Table 8.1, based on an Intergovernmental Panel on Climate Change (IPCC) report by Schlömer et al. (2014), provides a comparative indication of the equivalent amount in grams of CO_2 emissions per kilowatt hour of different technologies for supplying electricity by relying on different energy sources.

Note that actual values of GHG emissions from supplying electricity can diverge considerably from the values shown in Table 8.1 (see Schlömer et al., 2014, Table A.III) depending on a variety of factors.[2] Nevertheless, the general pattern is not substantially altered as a result.

Table 8.1   Median life-cycle emissions (g CO_2/kWh) of selected currently available commercial technologies for producing electricity

Method of electricity supply	Median gCO_2 eq/kWh
Coal (powdered)	820
Gas (combined cycle)	490
Solar PV in utility	48
Solar BV – rooftop	41
Geothermal	38
Concentrated solar	27
Hydropower	24
Nuclear	12
Wind offshore	12
Wind onshore	11

Source: Extracted from Schlömer et al. (2014, Table A.III.2, p. 1335).

Additional Observations on Cost Estimations and Life-cycle Factors

When considering the cost of supplying electricity by alternative means taking into account the life-cycle of projects for its supply, it is likely to be helpful to compare the following different components of costs:

- pre-production costs;
- annual overhead costs;
- annual variable costs as a function of output; and
- decommissioning costs.

The various available technologies for producing electricity differ significantly in relation to these components as is clear from the comparative analysis by Schlömer et al. (2014). However, these authors do not take account of pre-production costs. Only construction costs are considered and not, for example, planning costs, the costs of obtaining approvals and so on. In addition, decommissioning costs are left out of the equation and could be considerable in the case of nuclear. They argue that because the length of life of nuclear power plants on average is very long, their discounted decommissioning costs would be very low. This, however, is questionable from a social point of view because a zero or very low rate of discount may be called for.

Another interesting comparison is the length of the gestation period and the predicted length of lifetime of plant. Other things being equal, variations in these items affect returns. The gestation period is also important

Table 8.2   Average construction time and average lifetime of plant for supplying electricity using alternative technologies

Technology	Average construction time in years	Average life-time in years
Coal – pulverized	5	40
Gas – combined cycle	4	30
Biomass – dedicated	4.5	40
Geothermal	3	30
Hydropower	5	50
Nuclear	9	60
Concentrated solar power	2	20
Solar PV – rooftop	0	25
Solar PV – utility	0	25
Wind – onshore	1.5	25
Wind – offshore	1.5	25

Source: Extracted from Schlömer et al. (2014, Table A.III.1).

because it influences the rapidity with which different methods of supplying electricity can be employed to increase supply and change the composition of sources of supply.

Table 8.2 provides some comparative data which have been extracted from Schlömer et al. (2014). These data are for the average construction time for currently available commercial technologies for electricity supply and their average plant lifetime.

While one should not put too much store by the figures in Table 8.2, they are interesting from an indicative point of view. Note that the construction time for photovoltaic solar is not actually zero but only a small fraction of a year. Also construction time does not take full account of the gestation period and the speed with which capital equipment and labour are available for construction. In the case of nuclear, the manufacture of plant required for construction of a nuclear power plant can take a considerable amount of time. Furthermore, given the rapidity of technological progress, a long lifetime for a plant is not necessarily an economic advantage because of obsolescence.

Ability to Control Electricity Supply

A major issue which needs to be addressed in the supply of electricity by alternative means is their ability to respond to variations in the demand for electricity. Hydropower can be quite responsive but is limited by the availability of water and elevations suitable for electricity generation. The supply of energy using solar can be quite variable and there is no supply at night. Wind power can be available at night but the strength of the wind can change considerably. To some extent, these problems can be reduced by adopting means to store energy and, in the case of wind power, possibly by having some excess capacity. Nevertheless, given current technology, the means of contributing electricity to the grid needs to be diversified in order to allow for variations in its supply by solar and wind. Given current technologies, rising dependence on solar and wind increases the risks of disruption to electricity supplies. Consequently, the use of hydrocarbons (particularly gas) is likely to supply a significant but most likely a declining percentage of electricity requirements for some time to come.

The Private and Social Economic Costs of Supplying Electricity by Alternative Means

It is difficult to generalize about the relative commercial (private) cost of supplying electricity by alternative means because costs tend to be situational and estimates rely on a range of assumptions. Furthermore, due to technological progress, costs can alter rapidly. For instance, the cost of supplying electricity using photovoltaic cells has fallen dramatically due to technological progress in panel design and increased economies of scale in their manufacture. Moreover, it is possible that technological advances in storing solar power (for example, in molten salt) will reduce the costs of deriving electricity from this source even further. Despite these difficulties, Schlömer et al. (2014) provide some cost estimates of alternative means of supplying electricity.

They find, for example, taking into account the whole life-cycle of different technologies and ignoring subsidies and taxes, that electricity can be produced at a lower private cost by wind onshore, hydropower and geothermal than by the use of coal but that it is more costly to do this by means of solar. However, the economic situation is not static and comparative costs vary with the geographical locality in which the electricity is to be generated. Furthermore, no account is taken of differences in the external environmental costs attributable to different methods of generating electricity.

The nature of the private costs functions for generating electricity also varies significantly with the technological means used to generate it. For example, the fuel costs of producing electricity using hydrocarbons are positive and vary with the level of production, whereas wind power and solar radiation are not priced: their variable costs of energy use are zero. Nevertheless, if private land is used for these flow resources, it is possible that private landholders could obtain higher rental income (a recurring fixed cost) for allowing solar and wind installations on land that is more favourably placed to take advantage of these natural resources. However, this would not be a variable cost but an on-going overhead cost.

The social economic cost of supplying electricity by all the means available exceeds its private commercial costs, ignoring taxes, subsidies and other government distortions. The difference is highest for the use of coal when its whole life-cycle and production chain is considered. It is a major contributor to air pollution as well as water pollution. Air pollution occurs at most stages in its chain of use – in mining, in transport and in its combustion. The difference between social economic costs and private economic costs is substantially less for the use of wind and solar power but not zero. Consequently, public policies that favour the use of wind and solar power rather than coal are justifiable from a collective point of view, as is public support for research efforts to find economical and environmentally friendly ways of storing energy obtained from these sources.

Political Lobbying and Lock-in

Political interference and lobbying has a major influence on energy supplies. Those who benefit economically from the mining and use of hydrocarbons form a strong political lobby. This contributes to lock-in to these sources of energy supply. However, the socio-economic system is not completely locked in to this technology. 'Green' groups have supported the use of alternative cleaner technologies. As the adoption and development of these technologies have progressed, new political pressure groups of producers have come into existence to support the case for their increased further development.

8.6   ECONOMIC STRATEGIES TO ALTER FOSSIL FUEL USE IN ORDER TO LOWER GREENHOUSE GAS EMISSIONS

Several different economic strategies exist for reducing GHG emissions stemming from the consumption of electrical power. These include government support for the development of technologies that have low intensities of CO_2 emissions; public support for the development and adoption of technologies that increase the efficiency of power use and generation; public policies that increase the relative demand for power generated by sources having a low carbon footprint; and measures to reduce the rate of economic growth, for example, establishing a steady state economy. It is not possible here to consider and investigate the relative merits of all these different possible strategies. However, it ought to be noted that some energy-saving developments and some carbon-friendly substitution of energy sources are likely to be generated by the market mechanism itself. Furthermore, sometimes only minimal government intervention may be called for to support market trends, for instance requiring that the comparative energy requirements of different brands of major electrical appliances be made known to buyers on a common rating scale.

Here I'll just consider briefly the fiscal policy of imposing a Pigovian tax on the use of coal and other hydrocarbons in order to lower CO_2 emissions. I'll also include a few comments on the proposition that establishing a steady state economy is a desirable way to reduce GHG emissions.

Imposition of a Pigovian Tax on Electricity Generated by the Use of Coal and Other Hydrocarbons

A levy (sales tax) may be applied to sales of electricity generated by the use of coal, with a lower levy on the use of other hydrocarbons having lower intensities of CO_2 emissions. Ideally, the magnitude of this levy should vary according to the comparative level of the sum of all negative externalities associated with different forms of power generation. This Pigovian measure will reduce the competitiveness of the supply of electricity generated by hydrocarbons and encourage switching to alternative means of supply which are less environmentally damaging. Both supply-side and demand-side responses can be expected if this levy is imposed. If the levy is increased at a gradual rate, it will limit economic dislocation in the coal-dependent sector.

While, in theory, not as efficient as a tradeable permit scheme for limiting CO_2 emissions, such a levy is likely to have advantages from a public administration point of view. It would be less costly to administer than a tradeable permit scheme and easier to monitor. Currently, tradeable permit schemes have a wide degree of support among scholars as a method of regulating CO_2 emissions by power stations. These schemes can vary considerably in their nature and their income distribution consequences (Tisdell, 2009). Many other economic policy instruments can potentially be used to regulate GHG emissions, each of which have their advantages and disadvantages. They have been widely discussed in the available literature (see, for example, Tisdell, 2009, Ch. 7). However, the adoption of any economic policies to influence GHG emissions is heavily dependent on political forces.

The Strategy of Establishing a Steady State Economy

The desirability of establishing a steady state economy in order to limit GHG emissions and other adverse environmental effects of human activity has also been canvassed (see, for example, Ross et al., 2016). Politically, however, it is likely to be difficult (probably impossible) to implement such a policy for the reasons outlined in Chapter 5. Furthermore, the nature of such a policy depends on how it is specified. Some formulations assume ZPG but it is difficult to see how this will be achieved globally in the near future. Some are based on zero growth in aggregate economic production or even a reduction in this production. However, if this is achieved, the total accumulation of GHGs in the atmosphere can continue to rise (if hydrocarbons continue to be a major energy source) because the rate of these emissions may still exceed the absorption capacity of sinks. A change is still required in the way in which power is generated. There is a need to adopt technologies which have low GHG intensities.

Zero economic growth is expected to make it more difficult to alleviate poverty and is likely to add to social conflict about the distribution of income. For this reason and others (including those outlined in Chapter 5), modern societies are unable to escape from the aim of pursuing economic growth. Therefore, the most realistic course of action appears to be to continue research and to find and improve energy-supplying technologies which have a low environmental footprint.

Jackson and Webster (2016) provide a perceptive and thought-provoking update on the limits to growth debate. They appear to be in favour of policies for degrowth but also suggest that degrowth may already be occurring due to secular economic stagnation. However, they do not identify policies that might be politically accepted to achieve degrowth. In those countries where degrowth or economic stagnation has occurred in recent times, it has not been a chosen path. Furthermore, desperate political efforts are being made to reboot economic growth (for the reasons outlined in Chapter 5 of this book), even though these do not always bear fruit. The election of Donald Trump to the American presidency on the promise of a strong economic growth agenda supports the hypothesis that modern market economies are trapped in (and obligated to) continuing economic growth as a socially demanded economic attribute.

8.7   INCREASING SEQUESTRATION OF CO_2

A third possibility for combating global warming due to human activity is to adopt measures to increase the sequestration of CO_2 and avoid the loss of natural resources which store or are able to capture and store CO_2. In this regard, most attention has been given to the conservation of terrestrial organic biomass, particularly forests, and the possibility of increasing tree cover. However, the capacity of land-based plants to absorb CO_2 is limited and unless current use of fossil fuels is reduced dramatically, these plants will be unable to absorb the amount of CO_2 released. Furthermore, given current levels of world population and human demands for agricultural products for their sustenance, this severely limits the amount of land that can be forested, or the increase in tree densities which can be achieved, without causing a significant food shortage. Nevertheless, some increase in sequestration of CO_2 in organic matter is possible without adverse economic repercussions. Furthermore, there can be solid economic arguments for reducing loss of tree cover when all the environmental consequences of this are taken into account. However, one needs to question whether, on their own, measures to prevent a reduction in tree cover (or increase it) can be sufficient to reduce CO_2 accumulation in the atmosphere, given the volume of current and anticipated use of fossil fuels for some time to come.

A variety of economic measures are available to provide economic motivation for the conservation of sinks or to encourage additions to these. These include subsidies, taxes and prohibitions with penalties for non-compliance. Consider these aspects in relation to tree cover.

Subsidies for retaining tree cover (or adding to it) to sequester CO_2 have been classified as a form of payment for the supply of ecosystem services, and in some cases, have been coupled with offset policies. As with many types of payments for ecosystem services, their implementation can pose difficulties. First, payment ought (on allocative goods) only be made in those cases where, in the absence of the subsidy, an arboreal sink will disappear or be much reduced. Secondly, retention of CO_2 may only be temporary unless the payments are on-going. When they cease, tree cover may be reduced or destroyed, adding the previously stored CO_2 to the atmosphere. The third problem is that agency costs of ensuring compliance can be high (for example, monitoring and policing compliance) and this can give rise to moral hazard problems, for example, corruption. A fourth aspect is that merely providing economic incentives for supplying a single ecosystem service can have unbalanced economic and environmental consequences. For instance, plantation forest monoculture might be supported at the expense of natural plant cover on the grounds that the former is an effective sink for CO_2. However, other important ecosystem benefits may be lost, such as the preservation of unique wild species.

Each of the above aspects is relevant to the preservation of sinks (or addition to these) as offsets to those economic developments which add to CO_2 emissions to the atmosphere (for example, CO_2 emitted by new power stations using hydrocarbons). In addition, the adoption of such an offset policy might amount to little more than window-dressing because the amount of extra carbon stored may be much less than the amount generated by the economic activity it is intended to offset.

Taxes and/or administrative restrictions on the removal of trees (or other forms of vegetation) are alternative policies to the 'carrot' approach for maintaining the amount of CO_2 sequestered in botanical matter. In some jurisdictions, permits are required for tree-clearing on private land but the regulations applying to such clearing often alter as the make-up of governments in political power change. Land cleared of tree cover is rarely revegetated to achieve the former amount of tree cover. Consequently, with the passage of time, tree cover continues to decline in many nations, as it has done, for instance, in Australia. On the whole, agricultural lobby groups have opposed regulations that limit the clearing of land. This they have done mainly on the grounds that it restricts the ability of farmers to determine the use of their land, reducing their property rights in a natural resource. This often occurs without economic compensation to resource-holders and therefore is controversial (Seidl et al., 2002).

8.8 FURTHER DISCUSSION OF SUSTAINABILITY AND OTHER RELEVANT 'GREEN ISSUES' INVOLVED IN ENERGY PROVISION

Some confusion seems to exist among members of the general public about the relative environmental merits of using alternative sources of energy for power and heating. For example, some environmental groups supported the use of biofuels because they believed they constituted a renewable and sustainable source of energy. However, their use and that of biomass for power production and heating is problematic. Let us consider some aspects of this as well as similar issues for other sources of power and heating.

Use of Biomass and Biofuels

The use of biomass for power and heating accelerates the return of CO_2 to the atmosphere. This is a negative effect even when biomass waste is used for this purpose, for example, bagasse from sugar cane, wood waste. Many rural communities in developing countries still rely heavily on biomass for heating, including animal manure. In some cases, biomass-use adds to deforestation, and the use of animal manure for heating reduces its availability for fertilizer for crops and pastures. The smoke from burning such biomass can also cause respiratory problems. Even though botanical biomass is a renewable resource, its sustainable potential supply is insufficient to support the basic needs of the current world population for power and heating, particularly when the opportunity cost of using land for this purpose (for instance, the value of food production forgone) is taken into account. A similar constraint applies to the production of biofuels (Tisdell, 2015, pp. 94–5).

In many countries, policy measures are in place to support the production and use of biofuels. For example, in the United States generous subsidies are provided for this purpose. These help to support the income of grain growers. In Europe there are mandatory measures to elevate the use of biofuels. These policies tend to increase the extension and intensification of agriculture, thereby reducing CO_2 sequestration. In some cases, they also have a negative effect on the conservation of wild biodiversity, for instance by encouraging the expansion of oil palm production (Tisdell, 2011). In addition, these policies can increase the price of food, for example, by adding to the demand for grain (Tisdell, 2015, pp. 94–5).

Flow Resources as a Sustainable Source of Energy

Several types of flow resources are highly sustainable sources of energy and their future supply is independent of the current level of their use. This is clearly so in the case of solar, wind and tidal. This adds to the attraction of relying on these natural resources as a source of energy. Some forms of hydropower merely rely on the flow of streams and therefore are a flow source of energy. However, others rely on dams. In some circumstances (for example, reduced inflow to a dam during drought), current use may be at the expense of future use. Therefore, they are not pure flow resources.

Climate change can affect the supply of all these sources of energy in different locations, except possibly tidal. For example, changes in rainfall and snow cover will affect hydroelectricity supplies. An increase in cloudy days may reduce the efficiency of solar in some cases and wind patterns could alter as a result of climate change. However, the periods during which these changes occur will probably be longer than the commercial life of power projects relying on these energy sources. Therefore, it ought not to be difficult to adjust to these environmental changes.

Whether or not energy from nuclear fission should be regarded as a flow resource is debatable. However, it is sometimes classified as a renewable resource because radioactive decay of uranium occurs naturally but extremely slowly. Its rate of decay is accelerated in nuclear reactors but can produce energy for a very long period of time. Politically the use of nuclear reactors is limited by the fear of nuclear accidents and doubts about how safely nuclear waste can be disposed of. The mining of uranium and similar sources of nuclear energy can pollute water bodies and this is a further concern.

Extending the Social Assessment of Hydrocarbons as a Source of Energy and Heating

Assessing alternative forms of energy provision (such as electricity) for only a part of their production and distribution chain can be very misleading because it presents an incomplete picture of the sustainability of this provision, the economics involved and its environmental consequences. Although the comparative analysis by Schlömer et al. (2014) is helpful, it is incomplete because it does not consider the whole chain of the production and use of resources for energy generation and heating, and it does not take full account of all the environmental consequences of alternative means of supplying energy and heating. Extending this analysis to better assess the desirability of using hydrocarbons is especially important. Although this cannot be done here in detail, some relevant considerations can be mentioned.

At the present moment (2016) the sustainability (their finiteness) of the supply of hydrocarbons is not a major problem. Of greater concern are the environmental consequences of their use for energy production and heating. Their mining raises several environmental issues. In the case of open-cut coal mining, coal dust (and other dusts) pollute nearby neighbourhoods. This creates a health hazard for those living nearby as well as a cleaning problem, and the productivity of surrounding agricultural land is reduced. Water bodies are frequently polluted by acidic water being discharged from such mines and this has negative effects on biota. In the eyes of most people, the scenic quality of the landscape is ruined. Furthermore, once the mine is abandoned, restoration of the landscape often does not take place.

The transport of coal also gives rise to several negative environmental externalities. Coal dust blows away from stockpiles of coal, as well as during its loading and unloading for transport, and sometimes from wagons when it is transported by rail. Although measures can be taken to minimize these effects (for instance rail wagons can be covered or coal can be sprayed with water), these measures are often not taken. The transport of coal also depends on methods that contribute to CO_2 emissions.

In the case of oil, spills at sea can create major environmental problems, for example, as a result of the wreck of oil tankers and the 'blowout' of offshore facilities for extracting oil. Oil refining adds to GHG emissions as do most of the end-uses of refined oil. The transport of oil and gas by means of pipelines can be another contentious environmental issue. Coal seam gas recovery can also result in serious contamination of water bodies and have adverse consequences for underground aquifers.

A holistic assessment of the social (economic) desirability of alternative technologies and resources for supplying energy and heating must go well beyond assessing a part of their production and use chain. When this is done, both solar and wind power seem to be promising means for increased energy supply in the future, particularly if environmentally suitable means can be devised for storing the power they provide.[3]

8.9   CONCLUDING COMMENTS

For economic and political reasons there is likely to be growing use of solar and wind power to provide sources of energy and heating in the future. Nevertheless, fossil fuels are likely to remain a significant part of the energy mix for some time to come. Consequently, large amounts of GHG will continue to be contributed to the atmosphere and the concentration of CO_2 in the atmosphere will still rise, albeit possibly at a reduced absolute rate. In order to avoid the accumulation of CO_2 in the atmosphere and to avoid further global warming, it is not enough to reduce the intensity of CO_2 emissions or even the rate of absolute emissions as a function of time. Public policies and private decisions which achieve these results will not stop global warming but should result in it being less severe than otherwise and should moderate adjustment problems.[4]

The Paris Agreement on Climate Change (to which all countries agreed in 2016) is intended to limit the growth in CO_2 emissions. Each nation is to use its own discretion about how it will contribute to this goal, and as a follow-up, each is to be asked to outline how it intends to achieve this goal. After five years, each is to report on its success or otherwise in meeting its objectives. While the Paris Agreement might reduce the growth in anthropocentric CO_2 emissions, there will probably still be an increase in the absolute amount of these. Consequently, the world will not have escaped from further global warming and its environmental consequences (for example, sea level rises) nor from the likelihood of an increase in the acidification of the oceans. Nevertheless, if this agreement helps to moderate these environmental developments, it will have been worthwhile.

NOTES

1. Within each of these categories, there are often quite different methods of generating power. Useful general information about this can be found from the web. In this case, the Wikipedia entries are especially helpful.
2. For example, the amount of GHG emissions from dams used for generating hydropower is usually considerably higher closer to the tropics than in areas further away because of the difference in the rate of decay of organic matter flooded by dams.
3. Several significant sources of anthropocentric GHG emissions have not been covered in this chapter. These include emissions from transport and from industries, such as cement, iron and steel, chemicals and pulp and paper. Some information about the contribution of these to GHG emissions is available in Schlömer et al. (2014). Given the word limit for this book, I decided that it was best just to focus on electricity production in this chapter.
4. It had originally been intended to discuss adjustment issues as part of this chapter but, given the word limit for this book, this proved to be impractical. However, some attention is given to these issues in Chapters 9 and 10. Migration is likely to be a very important issue in this context as it has been in the past in response to climate change (Cunliffe, 2015). In many cases, international migration will be necessary for some communities, for example, for the residents of low-lying island nations (such as Tuvalu and Kiribati) likely to be completely flooded by sea level rise.

REFERENCES

Cunliffe, B. (2015), *By Steppe, Desert and Ocean*, Oxford and New York: Oxford University Press.

Earth's CO$_2$ Home Page (2016), 'Atmospheric CO$_2$ [today's date]', accessed 10 November 2016 at https://www.co2.earth.

Hobbes, T. (1651 [2010]), *Leviathan, or the Matter, Forme, and Power of a Commonwealth, Ecclesiasticall and Civill*, New Haven, CT: Yale University Press.

Jackson, T. and R. Webster (2016), 'Limits revisited: a review of the limits to growth debate', *APPG Limits to Growth*, accessed 8 December 2016 at http://limits2growth.org.uk/revisited/.

Pearce, D.W. (1998), *Economics and Environment: Essays on Ecological Economics and Sustainable Development*, Cheltenham, UK and Northampton, MA, USA: Edward Elgar Publishing.

Ross, L., K. Arrow, R. Cialdini, N. Diamond-Smith, J. Diamond, J. Dunne, M. Feldman, R. Horn, D. Kennedy, C. Murphy, D. Pirages, K. Smith, R. York and P. Ehrlich (2016), 'The climate change challenge and barriers to the exercise of foresight intelligence', *BioScience*, **66**(5), 363–70. DOI: 10.1093/biosci/biw025.

Schelling, T.C. (1995), 'Intergenerational discounting', *Energy Policy*, **23**(4), 395–401. DOI: 10.1016/0301-4215(95)90164-3.

Schlömer, S., T. Bruckner, L. Fulton, E. Hertwich, A. McKinnon, D. Perczyk, J. Roy, R. Schaeffer, R. Sims, P. Smith and R. Wiser (2014), 'Annex III: technology-specific cost and performance parameters', in O. Edenhofer, R. Pichs-Madruga, Y. Sokona, E. Farahani, S. Kadner, A. Seyboth, A. Adler, I. Baum, S. Brunner, P. Eickemeier, J. Savolainen, S. Schlömer, C. von Stechow, T. Zwickel and J.C. Minx (eds), *Climate Change 2014: Mitigation of Climate Change. Contribution of Working Group III to the Fifth Assessment Report of the Intergovernmental Panel on Climate Change*, Cambridge and New York: Cambridge University Press, pp. 1329–56, accessed 28 April 2017 at https://www.ipcc.ch/pdf/assessment-report/ar5/wg3/ipcc_wg3_ar5_annex-iii.pdf.

Seidl, I., C.A. Tisdell and S. Harrison (2002), 'Environmental regulation of land use and public compensation: principles, and Swiss and Australian examples', *Environment and Planning C: Government and Policy*, **20**(5), 699–716.

Stacey, F.D. and J.H. Hodgkinson (2013), *The Earth is a Cradle for Life: The Origin, Evolution and Future of the Environment*, Hackensack, NJ: World Scientific.

Svizzero, S. and C.A. Tisdell (2014), 'Theories about the commencement of agriculture in prehistoric societies: a critical evaluation', *Rivista di Storia Economica*, **30**(3), 255–80.

Svizzero, S. and C.A. Tisdell (2016), 'Economic evolution, diversity of societies and stages of economic development: a critique of theories applied to hunters and gatherers and their successors', *Cogent Economics & Finance*, **4**(1). DOI: 10.1080/23322039.2016.1161322.

Tisdell, C.A. (1968 [2015]), *The Theory of Price Uncertainty, Production and Profit*, Princeton, NJ: Princeton University Press.

Tisdell, C.A. (1989), 'Environmental conservation: economics, ecology and ethics', *Environmental Conservation*, **16**(2), 107–112.

Tisdell, C.A. (1990), *Natural Resources, Growth and Development*, New York, NY, Westport, CT, and London: Praeger.

Tisdell, C.A. (2009), *Resource and Environmental Economics: Modern Issues and Applications*, Singapore, Hackensack, NJ, London and Beijing: World Scientific.

Tisdell, C.A. (2011), 'The production of biofuels: welfare and environmental consequences for Asia', in M. Hossain and E. Selvanathan (eds), *Climate Change and Economic Growth in Asia*, Cheltenham, UK and Northampton, MA, USA: Edward Elgar Publishing, pp. 38–61.

Tisdell, C.A. (2015), *Sustaining Biodiversity and Ecosystem Functions: Economic Issues*, Cheltenham, UK and Northampton, MA, USA: Edward Elgar Publishing.

Tisdell, C.A. and S. Svizzero (2017), 'Optimization theories of the transition from foraging to agriculture: a critical assessment and proposed alternatives', *Social Evolution & History: Studies in the Evolution of Human Societies*, (in press).

9. Agriculture and environmental change, especially climate change: economic challenges

9.1 INTRODUCTION

The purpose of this chapter is initially to identify important environmental changes caused by agricultural activity and development and to consider why these may result in economic welfare being less than it need be. Agriculture's role in promoting economic growth and development and in environmental change is then placed in its long-term historical context. Subsequently, the challenges facing agriculture this century in terms of meeting expected increased demands for agricultural production and the goal of environmental preservation are examined. This is followed by consideration of the possible economic consequences for agriculture of global warming and an assessment of strategies to cope with it.

9.2 GENERAL RELATIONSHIPS BETWEEN AGRICULTURAL ACTIVITY, ENVIRONMENTAL CHANGE AND ECONOMIC WELFARE

Since agricultural development first commenced some 12000 to 10000 years ago in the Near East, it has contributed significantly to economic growth, has supported a huge increase in the global population of human beings and has arguably been a major anthropocentric source of environmental change. Economic demands on agriculture continue to increase. Considerably more food (especially animal-based food) will need to be produced in this century by agriculture to satisfy the demands of a rising global population as well as additional demands generated by higher levels of income in some parts of the world, for example, China. This raises the question of whether agriculture can be sustainably developed to meet this challenge without causing socially unacceptable environmental deterioration, particularly given the anticipated effects of global climate change.

Addressing this matter effectively depends in part on the adequacy of

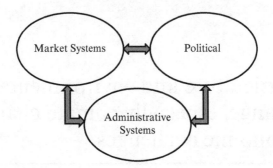

Figure 9.1   Several (partially interdependent) socio-economic systems influence resource use: shortcomings in any of these can result in failure to achieve desired social goals

existing socio-economic systems in managing resource use and the scope for altering these systems. In contemporary economies, resource use is basically organized by a set of three socio-economic factors:

- market forces;
- political forces; and
- administrative forces.

Some (but not complete) interdependence exists between these forces (see Figure 9.1). Their nature and relative importance varies as a rule between jurisdictions. Deficiencies in any of these subsystems for organizing resource use and economic activity can result in human well-being being less than it otherwise could be or in failure to achieve attainable collective goals.

Economists give the greatest attention to the ability of market systems to achieve social goals and to ways in which shortcomings in their operations can be rectified. They also pay attention to mechanisms to create markets by establishing private property rights. An example of the latter policy is the use of tradeable permits for the allocation of common pool resources, that is, cap and trade systems. These may be used, for instance, to regulate withdrawal of water by agriculturalists from shared water bodies such as underground aquifers or freshwater lakes. They are also being increasingly employed to control carbon dioxide emissions. However, there are also limits to the economic value of creating markets because transaction costs are associated with market operations. Furthermore, the creation and protection of private property rights can involve considerable costs. As has been pointed out by Daly (1999, Ch. 4), market-making is unable on its own to address adequately all of the environmental challenges we face.

In considering the effects of agriculture on the environment, it is convenient to divide these into on-farm and off-farm environmental changes. The latter are basically environmental spillovers or externalities. In many cases (but not all), they are a source of market failures because they result in adverse economic outcomes.

They may, for example, be judged to have socially adverse outcomes because the economic losses experienced by those unfavourably affected by the spillovers exceed the economic benefits of those generating these, or because this occurs beyond some level of intensity of the spillovers. Scope may then exist for socially beneficial regulation of these spillovers. However, if only on-farm environments are considered, (and environmental spillovers from agriculture are ignored) choices about land use may still raise social concerns.

Agriculture and On-farm Environmental Changes

The attributes of valuable on-farm natural resources for agricultural production are far from indestructible. This is especially true of soils. They are liable to decline in fertility and lose other valuable agricultural properties when used repeatedly for agricultural production, unless measures are taken to counteract these losses. The following are some possible adverse consequences of cultivation of soils:

- nutrient loss;
- unfavourable changes to soil structure;
- loss of soil depth;
- adverse chemical changes such as increased acidity of soils.

Furthermore, the grazing of land by livestock can result in unfavourable on-farm environmental consequences. These include:

- Accelerated soil erosion, especially as a result of overgrazing.
- Soil compaction.
- Changes in vegetation cover, for example, an increase (on rangelands) in the density of plant species which are unpalatable to or of low nutritional value for livestock. Consequently, the livestock carrying capacity of the land is reduced.

There is no doubt that agricultural land use can result in the unsustainability of agricultural yields. Agriculture is more susceptible to this problem in some areas than in others. It is, for example, a major problem for agricultural production on steep slopes having shallow soils, and less so on deep fertile soils on flat land. In many cases, lack of agricultural sustainability on individual farms can be counteracted by the use of off-farm inputs (such as fertilizer inputs) and by altering farming practices, such as engaging in crop rotation and changing cultivation methods. The extent to which this countervailing behaviour occurs is influenced by such factors as the availability of suitable off-farm inputs, the availability of techniques that can enhance the sustainability of agricultural productivity and the economics of taking advantage of their availability as well as the objectives of individual farmers.

Given the options available to farmers and considering only on-farm production, some are likely to adopt farming practices that result in less sustainability of yields than is desired by society as a whole. Their preference for economic returns may exhibit greater tilting towards the present than is socially desired. This would be reflected in their effective rate of discount of economic returns being higher than the social rate and/or in their planning horizon being shorter than is socially desired. Reasons for this could include:

- demands by farmers to accelerate their earning of income to provide support for offspring to move off-farm, for example, by financing education of their offspring, or to help finance their own exit from farming;
- excessive optimism about the prospects of new technologies being developed which will offset declining yields;
- ignorance; and
- insecure land tenure.

Environmental Externalities Arising from Farming

Environmental changes brought about by agriculture are not confined to individual farms. They also spill over to other farms, other parts of the economy, and to non-farmed environments. The geographical extent of these environmental spillovers varies. Some are localized, such as the salting of soils caused in some areas by tree removal. Others may have a wider geographical impact such as the nutrient-enrichment of marine areas containing coral reefs, as a result of discharges from major river systems. Those elevated levels of nutrient discharges may primarily be a result of fertilizer use and soil disturbance caused by agriculture in the catchments of river systems draining into offshore areas (see the next chapter). In some instances, the environmental effects of agriculture are of an even wider global nature, for example, they contribute to global climate change. In evaluating the environmental impacts of agricultural activity (and in devising policies to counteract any undesirable social effects of these), account should be taken of the geographical dimensions of the environmental spillovers involved. These are diverse in their nature and in their areal extent. In addition, they are liable to vary in other important respects, for instance, in whether they are reversible or are easily reversed, and in the timing of their environmental consequences. This can complicate the formulation of socially optimal policies for the control of environmental spillovers, if their control is warranted on economic grounds.

As the world's human population multiplies and as demands for extra agricultural production rise, concerns about environmental spillovers from agriculture (and associated sustainability issues) can be expected to increase. An increase both in the intensity of agricultural production and in the geographical extension of agricultural activity seems likely to occur to meet these demands. Consequently, environmental spillovers from agricultural activity can be expected to become more pervasive (a continuation of its historical trend) and will pose on-going policy problems.

Environmental spillovers from agriculture are very diverse. When they occur, private net benefits obtained as a result of agricultural decisions differ from their social net benefits. As a result, private decisions about resource use in agriculture are unlikely to be socially optimal, if social optimality is assessed using the standard economic assumptions employed in social cost–benefit analysis. Environmental spillovers can occur when agricultural land is private property, or if there is open access to it, or if its shared use is communally regulated. A local community may be guided in its land use only by its own self-interest and take little or no account of the environmental consequences of that for other communities.

Several adverse economic and environmental consequences have been noted in the literature for open access to land used for agriculture (mostly livestock grazing) in the area where it occurs. However, the economic and environmental footprint of such use is liable to spill over to other geographical locations. For instance, overgrazing on arid lands as a result of open access can increase the frequency and intensity of dust storms. Even communal regulation of shared land use for agriculture can result in adverse environmental spillovers. This is liable to occur if the relevant community is guided only by its own self-interests. Both the supply of public goods/bads and the aggregate availability of private commodities can be adversely affected by environmental spillovers from agricultural activity.

Table 9.1 provides examples of some adverse environmental activities which can arise from agricultural activity. However, it should also be realized that some agricultural developments and activities can give rise to positive spillovers. This has been claimed, for instance, for the terracing of sloping land for the growing of rice. Furthermore, it is not optimal from a social economic point of view (for example, if the Kaldor–Hicks criterion

Table 9.1   Examples of adverse environmental externalities from agricultural activities

Type of economic activity	Nature of possible environmental spillovers[a]
Tree removal	Water flows in river catchments become more erratic; soil erosion increased resulting in increased sedimentation of waterways and/or wind erosion; reduced storage of CO_2; in some areas, dryland salting of soils.
Cultivation of soils	Increased soil erosion; greater nutrient run-off; less CO_2 retention.
Fertilizer use	Increased nutrient run-off into water bodies. Several adverse environmental effects are possible from this. These include accelerated eutrophication of water bodies, loss of coral reefs and increased occurrence of blue-green algae invasions.
Pesticide use	Depending on the pesticide used, negative effects occur on some valuable species, such as pollinators (e.g. bees), and on the health of humans.
Heavy grazing of livestock	May result in reduced vegetation cover and in increased soil erosion. Change in the composition vegetation favouring some wild species but detrimental to others.
Excrement and waste from livestock husbandry, especially intensive husbandry	Increased nutrient run-off resulting in pollution of water bodies, offensive odours.
Habitat change	Loss of some wild species but others may be favoured. Overall wild biodiversity is decreased.
Irrigation using surface water	Waterlogging of low areas of land; salinity problems in some areas; adverse consequences for some wild species but some favoured.
Irrigation and withdrawal of underground water	In some areas, land subsidence; intrusion of saltwater.
Ponds for shrimp (prawn) farming	Salt intrusion into nearby rice fields, loss of mangroves and coastal wetlands.
Impact of exotic species for agriculture	Negative effects are possible on the conservation of some endemic wild species.

Note: [a] Nutrient and sediment loss from agricultural activities can be major contributors to eutrophication of water bodies, loss of coral cover, the occurrence of red tides and blue-green algal outbreaks as well as the presence and size of marine dead zones.

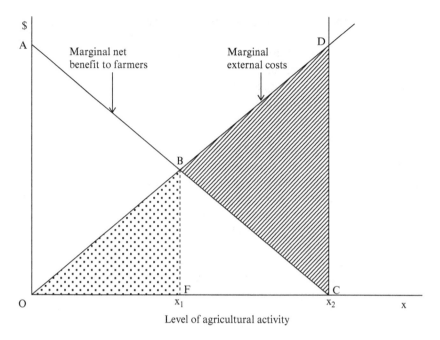

Figure 9.2 An illustration that it is usually not socially optimal to eliminate all external environmental costs associated with agricultural activity

is adopted) to eliminate all adverse environmental spillovers associated with agricultural activity.

Figure 9.2 illustrates the proposition that some level of adverse spillover from agricultural activity is often socially optimal from an economic point of view, for example, if the Kaldor–Hicks criterion is adopted. In this figure, line ABC represents the marginal net economic benefit obtained by farmers from agricultural activity and line OBD indicates the marginal external costs associated with it. If farmers are guided only by their own self-interest, they will engage in a level of agricultural activity of x_2. However, a level of activity of x_1 is socially optimal. The net social benefit of reducing the level of economic activity from level x_2 to x_1 is equivalent to the area of triangle BCD. Reducing the level of farming activity below x_1 would result in a greater loss to farmers than the increase in benefit to those adversely affected by the spillover. At the level of activity, x_1, an adverse externality still exists and this imposes environmental costs on those adversely affected by it. These costs are equivalent to the area of triangle OBF.

Of course, there can be situations in which the environmental costs associated with any level of agricultural activity exceed its private benefits and therefore, in such places, agriculture should not be permitted. Furthermore, the model in Figure 9.2 is a simple static single period one. Not only are the relationships shown in Figure 9.2 liable to shift with the passage of time but farming activity in one period is likely to have future economic and environmental consequences.

The fact too that agricultural activity is less than socially ideal when its environmental spillovers are taken into account does not mean that they can be corrected economically by policy interventions. This is because there are costs involved in the administration of policies to address the existence of environmental spillovers. These are liable to vary with the type of policy instruments used. When these costs and the effectiveness of available policies for controlling negative agricultural externalities are taken into account, it may be impossible to bring about a collective economic gain by a policy intervention.

Note that the model depicted in Figure 9.2 focuses on the flow of economic benefits and costs associated with agricultural activity. It does not take account of changes in the stock of various forms of capital, such as natural capital and heritage capital. For example, increased agricultural activity often leads to a socially unwanted deterioration in natural ecosystems and in a permanent and irreversible reduction in the value of services they supply. While sustaining all natural ecosystems in a pristine state is not an ideal choice from a collective economic point of view, a major challenge is to limit changes in these systems to ones that add to economic welfare and avoid changes that reduce it. This requires adequate attention to be given to trade-offs and opportunity costs as well as sustainability issues involving time patterns of economic and environmental change.

A case also exists for studying the historical development of agriculture and its socio-economic and environmental consequences rather than relying entirely on ahistorical models to assess these. As a result, a deeper appreciation can be gained of the role of agricultural activity and development in influencing economic growth and development, as well as environmental and social change.

9.3   AGRICULTURAL DEVELOPMENT IN HISTORICAL CONTEXT AND ITS ENVIRONMENTAL AND SOCIO-ECONOMIC CONSEQUENCES: A BRIEF ACCOUNT

Compared to the length of the time span in which *Homo sapiens* is believed to have existed, the cultivation of crops (the commencement of agriculture) is a relatively recent phenomenon. Agriculture is believed to have first commenced in the Near East about 10000 years ago and subsequently and independently in other parts of the world (Renfrew, 2013). Possibly, *Homo sapiens* emerged about 200000 years ago from a hominid ancestor (Renfrew, 2013, p. 79). Therefore, agriculture may only have existed for about 5 per cent of the duration of the existence of *Homo sapiens*.[1] At what time agricultural activity overtook hunting and gathering as the globally predominant means of obtaining human food supplies is unclear. In any case, the global shift from hunting and gathering as a means of providing supplies of food for humans was not instantaneous. The almost complete dependence of humans globally on agriculture (including aquaculture) to provide food appears to be a relatively recent phenomenon. What, however, is clear is that in the relatively short period of its existence, agriculture has enabled extraordinary economic growth to occur and has resulted in substantial environmental change, both directly by altering natural ecosystems and indirectly by providing essential support for those engaged in non-agricultural economic activities and developments.

The availability of food is a major constraint on the level of human population, and an agricultural food surplus is necessary to support non-agricultural production, for example, manufacturing and the supply of services. If there is an agricultural food surplus, it provides a basis for increasing economic specialization, the division of labour and the growth of cities. However, the agricultural surplus needs to be storable and to be able to be transported relatively easily to contribute significantly to these results. The development of agriculture was a crucial step in the evolution of economic systems. Childe (1936 [1966]) describes it as giving rise to the First Economic Revolution, the Industrial Revolution being the second one.

Without the presence of a suitable agricultural surplus and an increase in its size, insufficient food would have been available to support urban workers involved in manufacturing and other activities associated with the Industrial Revolution. In the absence of the development of agriculture, the Industrial Revolution would have made little progress.

The initial development of agriculture was very slow and limited to a few regions of the globe. Gradually, human dependence on agriculture (for survival) spread geographically and new techniques were also developed slowly. Consequently, global agricultural production increased and began to support an increasing level of population. However, it is probable that the standard of living of the bulk of the population did not improve compared to that of hunters and gatherers until after the Industrial Revolution. In fact, it may have declined. Most people seem to have survived at or near subsistence level.

Despite this, agriculture continued to develop, and urbanization increased. A possible explanation for this is that agricultural development changed the institutions and organizational structure of societies in places where it yielded a storable and transportable food surplus. This provided a basis for the emergence of central dominant elites who were able to appropriate this food surplus. This they could do over an expanded geographical area. These developments resulted in the birth of nation states (for example, initially palace-based economies) able to foster significant and long-lasting economic growth (Tisdell and Svizzero, 2015). The economic and social evolution of ancient economies is complex and I cannot provide an account of it here. Here I want to concentrate on the macro-environmental impact of agriculture.

Ehrlich (1989) has suggested that as a rough guide, the state of the natural environment depends on the level of human population (P), their per capita level of consumption of commodities (A), and the nature of technology (T) used for economic activity. Other things being held constant, 'deterioration' in the state of the natural environment tends to increase with P, A, and the use of technology that causes increased damage to (or changes in) the natural environment. In other words, the footprint of humankind on the natural environment tends to increase with P, A, and with the adoption of technologies impacting more adversely on the natural environment.

As a result of the advent of agriculture, the carrying capacity of the Earth for humans increased significantly (but at first slowly) compared to conditions in which humans depended entirely on hunting and gathering for their existence. While for several millennia, the development of agriculture seems not to have resulted in an increase in the standard of living of the masses, it was associated with a considerable increase in the global population, increasing wealth for a small proportion of the population (the dominant class), and the adoption of technologies resulting in accelerated changes in the natural environment compared to those previously brought about by the activities of hunters and gatherers.[2]

When agriculture commenced, it had little effect initially on the level of the global population. In all probability the food surplus it generated was small, agricultural techniques improved slowly and the geographical diffusion and adaptation of agricultural practices did not occur rapidly. Eventually agriculture was able to support an ever-increasing level of global population and contribute to economic growth, even though the masses remained extremely poor. The development of metallurgy resulted in its applications to agriculture and this raised its productivity and facilitated its geographical extension. The demand for metal products also provided a stimulus for increased interregional trade. There were also continuing improvements in methods of transport. These factors might help explain the rapid rise in global population after the Bronze Age commenced.

After the slow rate of global population growth between 1200 and 1650 AD, global population increase began once more to accelerate and was accompanied by rising agricultural production and global trade. European discoveries of new territories and their colonization played an important role in this process. At the same time, there were improvements in means of transport, in manufacturing and in agricultural techniques.

After 1750, the increase in the level of global population was very rapid. The mid-eighteenth century is often believed to mark approximately the beginning of the Industrial Revolution. Haub (2011) reports that in 1750, the level of global population was 795 million, rose to 1.656 billion in 1900 and reached 5.76 billion in 1995. By 2015, the level of global population exceeded 7 billion. The Industrial Revolution and increased globalization in its early stages stimulated global population growth. Subsequently, however, it resulted in a reduction in the rate of population growth as demographic transition proceeded in those countries that successfully industrialized and experienced significant economic growth.[3] Despite this, the level of global population continues to increase. It is predicted to increase to over 9 billion by 2060, and before the end of this century it may reach more than 10 billion.

Returning to the simple relationship between economic development and its impact on the natural environment suggested by Ehrlich (1989), the evolution of agriculture in ancient times was associated with increased population as well as the development of techniques that facilitated the transformation of the natural environment to make the environment more suitable for agricultural production. Rising per capita income was not, it seems, associated with this type of economic development. However, considerable increase in inequality in the distribution of income occurred. In many cases, the elite siphoned off the agricultural surplus and used it to promote the accumulation of man-made capital and to foster other economic changes, which added to their wealth (Svizzero and Tisdell, 2014). In turn, this inequality factor (not highlighted in Ehrlich's relationship) also contributed to environmental change because it enabled investments to be made in altering the natural environment to facilitate economic production.

After the Industrial Revolution, the impact of economic growth on the natural environment accelerated markedly. This period saw increased population levels, increasing levels of per capita consumption, higher levels of accumulation of man-made capital and the development of techniques making it easier to transform the natural environment. Agricultural production grew rapidly to support a rising global population both as a result of its extension and its intensification. Extension of agriculture occurred in many parts of the world, including the Americas, parts of Africa and Australia and New Zealand. While extension of agriculture is still occurring, it seems now that increases in agricultural production have become more reliant on intensification than on extension. Increased agricultural yields have become increasingly dependent on genetic improvements in farmed organisms, the supply of off-farm inputs (such as fertilizers, pesticides, seeds, water for irrigation) and improved management of resources on farms. Given anticipated increases in global population in this century, the question has arisen regarding whether agriculture will be able to satisfy the demands of this population for food, and whether it can do this sustainably and in a manner that does not result in socially unacceptable environmental change. Taking into account its environmental effects, what is the best way for agriculture to respond to this challenge? Should the focus be on 'sustained' agricultural intensification or alternative policies to close the potentially emerging gap between demand for agricultural products and their supply?

9.4 COPING WITH INCREASED DEMAND FOR AGRICULTURAL PRODUCE IN THIS CENTURY IN A SUSTAINABLE MANNER AND WITHOUT UNDESIRABLE ENVIRONMENTAL CONSEQUENCES

A major challenge in this century is how to cope sustainably with anticipated increased demand for agricultural produce and avoid unwanted damage to the natural environment. The United Nations Department of Economic and Social Affairs (2012) predicts that the global human population will reach 9.5 billion by 2060. This is an increase of 2.3 billion (32 per cent) on its level of 7.2 billion in 2013.

The potential gap between current agricultural production and the increased demand for agricultural production could be met in a variety of ways. While most attention appears to have been given to ways of increasing agricultural production to fill this potential gap (that is, supply-side considerations), there is also the possibility of reducing the demand for agricultural produce, for example, by using less of it to manufacture biofuel. In addition, reducing post-harvest waste and possibly moderating demand for some high protein animal products, such as beef, could help.

Without a change in the distribution of the use and composition of agricultural produce, agricultural production needs to increase by about 30 per cent by 2060 from its current level to maintain the current status quo.

However, if the proportion of the world's population achieving increased per capita incomes rises (as is quite possible), this will add to the demand for increased agricultural production. In particular it is believed that per capita incomes in China will continue to rise in the foreseeable future. One high-end prediction (Burney et al., 2010) is that the demand for agricultural produce will rise by as much as 70 per cent by the middle of this century compared to its level early in this century. In any case, while this estimate appears to be rather high, an increase of more than 30 per cent in agricultural production is likely to be demanded by 2060. The question has arisen regarding whether this is likely to be achieved and how it might be achieved without increased damage to the natural environment. A commonly held view at present is that this can be achieved by the sustainable intensification of agriculture by adopting more efficient technologies. Let us consider this point of view.

Sustainable Intensification of Agriculture as a Means to Meet Increasing Demands for Agricultural Production in this Century

Ringler et al. (2014) suggest that by 2050 increases of about 20 per cent in maize, rice and wheat yields are possible as a result of sustainable intensification. The types of technologies they suggest be adopted to achieve this are listed in Table 9.2 in declining order of the percentage expected increase in yields. For example, they claim that the widespread adoption of no-till technology could raise maize and wheat yield by around 16 per cent, and more efficient nitrogen use could increase rice yields by about 20 per cent and reduce negative environmental spillovers from these crops.

The concept of the sustainable intensification of agriculture is appealing

Table 9.2   The set of agricultural technologies listed by Ringler et al. (2014) which, if adopted, are projected by their chosen modelling to increase maize, rice and wheat yields significantly by 2050[a]

1. Increased nitrogen-efficiency	7. Improved crop protection against insects
2. No till[b]	
3. Heat tolerant crop varieties	8. Drought tolerance
4. Precision agriculture	9. Drip irrigation
5. Integrated soil fertility management	10. Water harvesting
6. Improved protection against weeds	11. Sprinkler irrigation (rather than flood irrigation)

Note: [a] Lower numbers correspond to higher percentage effects on yields; [b] Does not apply to rice.

because it promises a win–win situation, namely reduced resource use (particularly land) by farmers, greater yields and higher returns accompanied by superior environmental outcomes. According to Ringler et al. (2014, p. 43):

> By sustainably increasing agricultural production, we can enhance food availability, which is a necessary condition to combat hunger. Experts agree that greater production must be achieved by increasing yields while using fewer resources and minimizing or reversing negative environmental impacts. This sustainable agricultural intensification approach is fundamentally about making the current agricultural system more efficient through the use of new technologies or by improving current production systems.

While a 20 per cent increase in yields for maize, rice and wheat would help to fill the predicted gap between the demand for agricultural produce expected by the middle of this century, it would still fall short of the requirement of at least a 30 per cent increase in agricultural production to sustain current levels of per capita consumption of agricultural produce. A consequence could be that the poor will be increasingly deprived of food, particularly if the proportion of the world's population obtaining higher incomes increases and they substantially elevate their demand for animal protein.

It has also been claimed that from the point of view of conserving wild biodiversity, increasing agricultural production by its intensification is more desirable than by its extension (Balmford et al., 2012; Phalan et al., 2011a; 2011b). However, it seems unwise to generalize about this. Both agricultural intensification and extension can result in loss of wild species, usually different sets of wild species. For example, Bentham et al. (2003) and Tscharntke et al. (2012) identify ways in which agricultural intensification is likely to result in loss of wild species. This and related environmental issues are discussed in Tisdell (2015, Chs 5 and 8).

Even if agricultural extension does not occur as the demand for agricultural produce increases, account needs to be taken of the fact that the current intensity of use of land for agriculture varies considerably. Some is used for the low intensity grazing of livestock (for example, for cattle production in the semi-arid zone of Australia), whereas high intensity multiple cropping occurs in some parts of the world. Efforts are being made in some parts of Australia's semi-arid zone to grow crops using underground aquifers. This could result in significant environmental change and could result in the unsustainable withdrawal of water from aquifers. In reality, the desirability of intensifying agriculture (even if higher yields are sustainable) is context and location specific (Garnett et al., 2013).

In practice, there is absolutely no guarantee that a socially optimal combination of the intensification and extension of agriculture will take place even if it is appropriately identified. Market, political and administrative failures are likely to prevent this from happening.

Alternative Means of Closing the Emerging Potential Gap between the Supply of Food and the Demand for it

The approach of encouraging sustainable agricultural intensification in order to meet the increased demand for agricultural produce focuses on raising agricultural production without causing added harm to the natural environment. However, it is not the only possible strategy which could be adopted to deal with the challenge of increased demand for agricultural produce and limiting the possible adverse environmental consequences of responding to it. Some demand-side policies could help address this issue. These include:

- Reducing the use of agricultural land for growing crops to produce biofuel. If high subsidies for biofuel production and mandatory requirements in some countries to use biofuels were withdrawn, this would make more agricultural land available for food production.
- Lowering post-harvest food waste would also have beneficial results. Much avoidable food waste occurs in homes and elsewhere in higher-income countries.
- In individual cases where the consumption of animal protein is high, especially red meat consumption (for example, beef), scope exists for beneficially reducing its consumption and substituting the consumption of more plant protein for it. This could have beneficial health consequences and reduce the use of land for growing crops to feed livestock.

To some extent, changes in food prices could encourage the above-mentioned demand-side changes. If a gap between the supply of food and demand for it emerges globally, this can be expected to result in a rise in the real price of food. Moreover, it is possible that the real price of animal-derived protein (especially beef) will rise at a faster rate than the demand for plant-based food. Rising food prices (relative to biofuel prices) will increase the economic attractiveness of using agricultural land to supply food and provide an extra incentive to avoid post-harvest waste. An increase in the real price of food obtained from livestock should moderate demand for this type of food. Nevertheless, the price mechanism may not be as effective in this regard as desired. Furthermore, a rise in the real price of food will have negative welfare consequences for the poor.

9.5　GLOBAL CLIMATE CHANGE AND AGRICULTURE – BASIC IMPACTS AND MODELLING OF THESE

Apart from the challenge facing agriculture in this century because of an anticipated rise in global population and an increased proportion of the world's population having higher incomes, we are faced with agricultural production and adjustment problems due to global climate change. Although agricultural yields have been increasingly shielded from uncontrolled environmental influences (such as prevailing climatic conditions) by technical progress and the increased availability of external inputs (such as irrigation water and fertilizers), they are still subject to considerable variation as a result of alterations in these uncontrolled conditions. Global warming is expected to alter uncontrolled environmental conditions substantially in most parts of the globe, and affect agricultural yields.

Environmental Changes due to Climate Change and their Effects on Agricultural Production

Table 9.3 lists some anticipated environmental changes due to climate change and their consequences for agricultural production. Many of these effects will differ depending on the geographical location of agricultural land and most will require changes in agricultural practices. In some areas, agriculture may need to be abandoned, for example, in areas where water availability is reduced so much that agriculture is no longer viable or in places where agricultural land is flooded due to sea level rise. In other areas, it may be necessary to change the type of crops grown or change from the growing of crops to the grazing of livestock if an area becomes semi-arid.

While the economic value of agricultural production in some areas will decline as a result of climate change (in some cases, substantially), in other areas it is likely to increase due to more favourable weather patterns. This makes it difficult to predict the aggregate global change in agricultural production and in its economic value. This difficulty is compounded by uncertainty about changes in weather patterns.

Different Approaches to Measuring the Impacts of Global Warming on Agriculture

Mendelsohn and Dinar (1999) identify three different basic types of approach to measuring the impacts of global warming. These are:

*Table 9.3   A list of expected environmental changes due to climate change
and their consequences for agricultural production*

Environmental Changes	Agricultural Consequences
Higher temperatures	Livestock subject to increased heat stress – can reduce their productivity. Pollen sterility and failure of some crops to set seed. For example, wheat is susceptible.
Reduced water availability in some areas	Likely to result in lower levels of agricultural production in these areas and changes in agricultural land use.
Longer periods of drought with greater frequency in some areas	Consequences as above.
Greater weather extremes	Increased crop damage, for instance due to storms.
Higher levels of CO_2 in the atmosphere	This results in carbon fertilization and promotes the growth of some plants, for example, sorghum. However, the plants may be less nutritious and carbon fertilization also favours the growth of important weeds.
Sea level rise	Loss of agricultural land due to inundation, movement inland of salinity in coastal rivers and streams, salt intrusion into some freshwater aquifers, reduced agricultural production.

Source: Partially based on information in Ziska (2011).

- agronomic-economic models;
- agroecological zone analysis; and
- Ricardian cross-sectional analysis.

Mendelsohn and Dinar (1999) favour the Ricardian cross-sectional approach on the grounds that it allows more adequately than the other approaches for the adjustment of farmers to climate change. Agronomic-economic models draw on experimental data to predict changes in crop yields due to changes such as temperature, precipitation and atmospheric carbon dioxide given the existing spatial distribution of crops and no changes in adaptation (Mendelsohn and Dinar, 1999). This is basically a simulation approach to determining variation in yields. Although according to Mendelsohn and Dinar, farming methods are not permitted to vary, there is no reason why experiments cannot be done for changed farming methods, for example, for the delayed planting of some crops to avoid heat stress. Once changes in yield are estimated, changed levels of aggregate production, prices and net revenue can then be predicted.

Agroecological zone analysis is based on actual observations of how crop varieties and their yields vary by climatic zones. These data are then used to predict changes in the locations where different types of crops are grown as climate alters and how their yields are expected to vary. These yield changes can then be used to predict overall supply and market effects. A variation on this approach is to collect data on how annual yields alter with changed annual weather conditions and use the data in the same way. Different assumptions can be made about the spatial movement of crops and different combinations of climate change variables can be allowed for. Mendelsohn and Dinar (1999, p. 282) state: 'This approach is subject to the same limitations as the agroeconomic models that researchers must explicitly account for adaptation'. These authors claim that the cross-sectional Ricardian approach is better able to allow for farmers' adaptations than other approaches.

In an updated review of the modelling of the impacts of climate change on agriculture, Mendelsohn and Dinar (2009) add economic farm simulation models and computable general equilibrium (CGE) models to their list of methods used to measure the economic response of agriculture to climate change. The former approach basically estimates alterations in farm production functions likely to occur as a result of climate change and considers the response of farmers assuming that they are profit-maximizers. CGE models allow endogenously for market interdependence, which partial ones do not. However, CGE models usually rely on comparative static analysis and require considerable abstraction to make them tractable. Furthermore, the way in which 'shocks' are introduced into these CGE models is often unavoidably crude.

Mendelsohn and Dinar contend that the use of cross-sectional or panel analysis of net revenues or land values (dubbed the Ricardian method) is the superior approach to predicting agricultural responses to climate change because it is based on the 'revealed' production choices of farmers in relation to climatic differences. They assume that if the climate (environmental conditions) in a geographical area I alters so that it becomes similar to that which prevailed in area II, farmers in I will make similar production choices to those which were made in II. Spatial transfer of agricultural production in response to climate change is assumed. Figure 9.3 provides a simple illustration of the logic behind this approach.

Suppose that in area I farmers typically grow wheat and in area II sorghum. The net return per hectare from growing these alternative crops

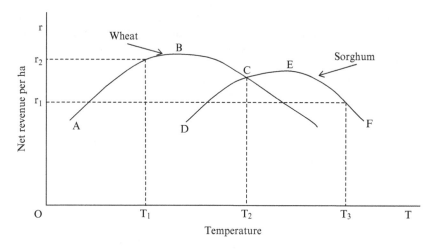

Note: This diagram is used to demonstrate the Ricardian approach to assuming the effects on agriculture of climate change.

Figure 9.3   *An illustration of net agricultural revenue per ha in relation to climate conditions proxied by temperature*

as a function of climatic conditions, proxied by the level of temperature, T, (after allowing for profit-maximizing adjustments by farmers) is as shown respectively by curves ABC and DCF. If farmers are profit-maximizing, they will switch from growing wheat to growing sorghum if temperatures rise above T_2. The envelope, ABCEF, is relevant to their decision-making. For example, if initially the relevant temperature in area II is T_1, wheat is grown and the average return per ha is r_2. If the relevant temperature is T_3 in area II, sorghum is grown and the average return per ha is r_1. Now suppose temperatures in area I rise to T_3. Farmers in area I would then switch to growing sorghum because this is their most profitable alternative. The switching point corresponds to a temperature of T_2.

The objective of this type of Ricardian analysis is to obtain, relying on empirical data, a mathematical approximation to the envelope of the type shown in Figure 9.3 and then use this for prediction purposes. The form of this envelope function is likely to vary from region to region. For example, it may be basically (1) increasing, (2) decreasing, or (3) rising at first and then declining in relation to altered climatic conditions. In the first case, climate change leads to an increase in the net value of agricultural production, a decline in the second case, and in the third case, it can lead to an increase or a decrease, depending on the magnitude (and nature) of climate change.

The above approach is a highly simplified one. For one thing, several environmental variables are likely to alter with climate change. These can include temperature, precipitation, evaporation and the time patterns of these. Agricultural production is sensitive to all these factors. In some cases, the changes will be highly correlated (for instance, rising temperatures may correlate with decreasing rainfall and an elevated evaporation rate), but not in all areas.

9.6 ALTERNATIVE VIEWS ABOUT AGRICULTURAL ADJUSTMENT AND NEGLECTED EFFECTS OF CLIMATE CHANGE ON AGRICULTURE

Mendelsohn and Dinar (2009, p. 309) believe that agricultural land use adjustments due to climate change are likely (in aggregate) to have benign global impact on agricultural supplies. They suggest that climate change may even have a positive effect on the net value of global agricultural production. Despite this, the quality of the available evidence seems to be inadequate to conclude this. Even if it were so, it should not be a reason for complacency because important distributional changes and adjustment problems will arise as a result of the effects of climate change on agriculture. These seem already to be occurring in some low rainfall areas, for example, in parts of Australia.

The following matters (which will now be considered) are worthy of increased attention: regional variations in the economic impact of climate change on agriculture; economic adjustment issues; adjustment implications of different forms of biological tolerance functions for alternative types of agriculture; likely changes in the economic structure of agriculture as a result of climate change; and implications of the movement of temperate crops towards the poles as a result of climate change.

Regional Variations in the Economic Impact of Climate Change on Agriculture

The adverse consequences of predicted climate change on agriculture are going to be severe in some parts of the world. In several cases, agriculture will have to be abandoned completely. Rises in sea levels will, for instance, result in some fertile land being flooded by the sea. Areas affected will include low-lying land adjoining the Bay of Bengal in Bangladesh and India, the Nile Delta and the lower Mississippi region. Other areas may become so arid that desertification takes place, making them unsuitable for any form of agriculture.

In other cases, the availability of water for irrigation is liable to shrink markedly and its supply could become more erratic. A considerable portion of Australian agriculture production relies on irrigation using water from the Murray–Darling river system. Climate change may significantly reduce average water inflows from the Darling River and its tributaries because of reduced rainfall and greater water evaporation in its headwaters, and also substantially (but to a lesser extent) in other parts of the Murray–Darling Basin (see Adamson et al., 2009, especially Table 2). Furthermore, these inflows are expected to become more erratic. In other parts of the world, reduced snowfall in the headwaters of streams used for irrigation will also impact adversely on the scope for irrigation. This is a concern, for example, in California. Reduced snowfalls (and accelerated snow melting) in the Sierra Nevadas could reduce the agricultural production of California substantially.

Even when farming is not abandoned, climate change may result in a change from high-value land use by agriculture to one of lower value, and consequently could negatively affect local economies that rely heavily on agriculture for their prosperity. This is not to suggest that agricultural production in all regions will be adversely affected by climate change; some will gain and some (but not all) nations may experience a net gain in the value of their agricultural production. However, to rely only on estimates of aggregate changes in agricultural production (a variant of the Kaldor–Hicks or potential Paretian improvement criterion) to claim that the effect of climate change on land use will be benign (favourable), as Mendelsohn and Dinar do, ignores distributional and adjustment issues of considerable importance for social welfare.

Some Neglected Adjustment Issues

Another limitation of the approach used by Mendelsohn and Dinar to assess the impacts of climate change on agriculture is that it is based on comparative static analysis. Consequently, it gives little attention to adjustment issues and the dynamics involved in adjustment. A problem faced by many regional communities is to know when their climate has changed permanently and the time-path of any change. This is important for investment decisions and relevant to whether and when farmers should alter their farming strategies. Uncertainty about this is likely to add to economic costs. Problems can include premature adjustment, or adjustment that is too much delayed, for example, changing types of crops grown too early or too late in response to permanent rises in temperature and reduced rainfall.

Implications for Agricultural Adjustment of Different Forms of the Biological Tolerance Function

Biological tolerance functions show how the biomass (or similar attributes) of species (including variety of species) alter with environmental conditions. These relationships can be complex and often depend on multiple variables. Frequently they are, however, just specified for variations in one environmental variable, all other environmental variables being held constant. Sometimes, they are depicted as bell-shaped (Tisdell, 1983) or as being a quadratic type of function with a single maximum (Mendelsohn and Dinar, 1999) and therefore of a symmetric nature. However, empirical evidence indicates that for many types of crops, they are not of a symmetric nature. For example, yields for many types of crops rise smoothly in relation to relevant prevailing temperatures but when some increased temperature is reached, they then fall sharply (Ziska, 2011). This is because high temperatures interfere with the flowering of the crop and the setting of seed.

The tolerance of different types of crops to heat stress varies. Sorghum is more tolerant than maize and maize is more tolerant than wheat. Consequently, a biological tolerance function of the form shown in Figure 9.4 occurs for most cereal crops. This means that once relevant temperatures exceed some critical value, yields from a particular type of crop drop

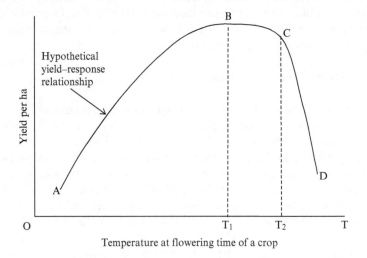

Figure 9.4   An example of a biological tolerance function that declines sharply once environmental conditions reach a particular threshold

sharply. Farm incomes will also do this unless satisfactory adjustment can be made in farming practices. In some cases (but not all) it is possible to alter planting dates of crops to avoid high temperatures while the crop is flowering or the farmer may switch to a crop which is more heat tolerant.

Is Smooth Economic Transition from One Form of Agriculture to Another Possible as Climate Changes?

The type of envelope analysis adopted by Mendelsohn and Dinar implies that the economic transition of farmers from one form of agriculture to another in response to climate change is relatively smooth. However, the relevant envelope of possible net returns may be slightly scalloped (wave-like) both for biological reasons and for economic reasons (see Figure 9.3), and in some cases, the troughs could be deep and wide. It may not, for example, be economic to switch from growing one crop to another until yields from the former fall substantially. Farmers may also have to learn about growing a new crop, so initially this may depress their returns when switching to it, because learning-by-doing is usually involved. Furthermore, new capital expenditure is likely to be required.

Structural and Institutional Change in Agriculture in Response to Climate Change

Possibly in developed countries, climate change will reinforce the trend favouring farm holdings of a larger size and the holding of these by public corporations (companies) rather than their ownership by individuals and families. In areas where climate change results in yields per ha falling, incomes from smaller-sized farms may fall to a level which is unacceptable to their owners. Thus, there is an incentive to amalgamate properties to ensure an adequate level of income.

Climate change increases agricultural risk in many areas. One way to reduce these risks is by means of the public company form of ownership involving limited liability. Furthermore, large companies can purchase land in different ecological zones, thereby reducing their risks even further, if returns are not perfectly correlated. An interesting variant of this is the holding by large Australian cattle enterprises of land in different ecological zones of the country. This enables (among other things) cattle to be moved from areas experiencing drought to properties where this is not so or to places where they can be lot-fed relatively easily.

Comment on the Movement of Temperate Crops Towards the Poles

As a result of climate change, the growing of temperate crops will move towards the poles and retreat from areas closer to the equator. While this might result in increased production of temperate crops in the Northern Hemisphere, it is likely to have the opposite effect in the Southern Hemisphere. This is because the total land mass in the Southern Hemisphere is significantly reduced as the South Pole is approached but this is not the case when the North Pole is approached.

9.7   BRIEF NOTES ON AGRICULTURAL POLICY AND CLIMATE CHANGE

Assuming that climate change is bound to occur, what agricultural policies (if any) should governments adopt to respond to it? Is, for example, a laissez-faire approach appropriate? This approach supposes that existing institutions are adequate to cope with climate change adjustments. To some extent, it is true that existing institutions can be expected to facilitate agricultural adjustments to climate change.

For instance, some farmers may find it worthwhile to make greater use of private crop insurance as the weather becomes more variable and uncertain. However, this will add to their costs of supply and because of the relatively high level of transaction costs involved, small-scale farms may find this type of insurance to be uneconomic. (See Tisdell et al. (2012) for a detailed coverage of the scope for private insurance and aquaculture. This coverage is also relevant to agriculture.) In cases where such insurance is available, it can smooth agricultural transition in response to climate change. However, private crop insurance (or insurance generally) is not a panacea for coping with climate change (Collier et al., 2008; Garrido et al., 2011). Furthermore, as mentioned above, business risks associated with climate change may also be reduced by the formation of agricultural public companies.

Most farmers today, particularly in developed countries, depend on inputs supplied by non-farming agribusinesses. Seed suppliers are likely to develop new seed varieties better suited to changed climate conditions. They may do this by genetic engineering, developing new hybrids, genetic selection and by drawing on seed banks. However, there are also limits to what can be done. For example, what can be achieved is likely to be limited by the stock of agricultural genetic material available. New techniques may also be developed by large non-farming agribusinesses to better enable farmers to cope with climate change. A balanced overview of possible technological agricultural innovations which might occur to cope with climate changes is available in Gollin (2011).

Income distribution problems are sure to arise as a result of climate change. Governments should take care not to prop up, by means of long-term subsidies, farms and farming practices which become uneconomic as a result of climate change. In some cases, the appropriate economic policy is to provide subsidies to farmers to move off farms that are no longer economic and support amalgamation of farms where this is needed to ensure their economic viability.

Unsustainable long-term changes in natural resource use to cope with climate change should be discouraged. For example, these include increased withdrawal of water from underground aquifers in response to reduced rainfall. This can result in water use exceeding the rate of recharge and a misallocation of water supplies.

Governments can make a worthwhile economic contribution to agricultural adjustment to climate change by increasing the supply of public goods relevant to the adjustment. For example, improved weather forecasting is likely to be valuable in that regard. However, there are other types of information, knowledge or know-how which are public goods (or virtually public goods), which might be beneficially provided by governments. These goods will either not be supplied by private enterprises or, if they are, their supply will be less than socially optimal. A problem, however, is that one cannot always be sure of the accuracy of information provided by public bodies.

9.8   CONCLUDING COMMENTS AND RESUMÉ

Agriculture is both a major source of environmental change (several examples of its external environmental effects were given in Table 9.1) and is itself quite sensitive to environmental change. Although technological change and economic developments have reduced the sensitivity of agriculture to environmental changes, it still remains sensitive to these changes, for example, to climate change. Furthermore, since the stock of heritage genetic agricultural material has been much reduced, as well as the stock of wild genetic material potentially of use in agriculture, this may add to the difficulties of adjusting agriculture to climate change.

Both on-farm and off-farm environmental changes due to agricultural activity were discussed. Explanations were given of how and why these changes can result in greater (avoidable) economic scarcity than is socially desirable. It was pointed out that market political and administrative failures can give rise to this result. Even though agriculture has been practised

for only a short recent period in the existence of *Homo sapiens*, it has arguably been the source of major and most pervasive sources of environmental changes due to human activities. For most of human existence, hunting
and gathering was the only means of human sustenance, and remained
the globally dominant source of human provisioning for a considerable
time after agriculture began. Agricultural development eventually made
possible a large increase in the human population and the establishment
of cities. It enabled a greater supply of food and other provisions to be
achieved than could be obtained by relying on hunting and gathering.
Eventually, agriculture generated a sufficient food surplus of a type which
played a major role in economic growth and which provided essential
support to those directly involved in the Industrial Revolution. Thus,
indirectly, agriculture supported economic changes outside of agriculture, which also brought about massive environmental change. Feedback
mechanisms from the industrial and service sectors for which agriculture
provided essential support magnified environmental changes caused by
human economic activity.

Agriculture still plays a vital role in the process of economic development and it continues to cause significant environmental change and to
be sensitive to it. In this century, agriculture faces major economic and
environmental challenges. Two of the most important ones discussed in
this chapter are:

- How will agriculture cope with a substantial increase in the demand
  for agricultural produce in this century without causing socially
  unacceptable environmental damage and can it cope?
- What will be the impact on agriculture of climate change and how
  will it cope?

In considering the first challenge, particular attention was paid to the prevailing belief that the most desirable way of coping with this anticipated
global increase in demand for agricultural supplies is to intensify agricultural production sustainably. Doubts were expressed about whether such
an approach is likely to be environmentally benign, as suggested by some
of its proponents. Alternative means (which can also be supplementary
ones and which would be more environmentally friendly) were suggested
for responding to the potentially emerging gap between the global supply
of food and the demand for it. These issues are given much more coverage
in Tisdell (2015, Chs 5 and 8).

Concerning agriculture and environmental change, several types of
impact of climate change on agriculture were identified (Table 9.2) and particular attention was given to different methods that have been employed

to measure the impacts of global warming on agriculture. The main emphasis was on the Ricardian model, which allows for the adaptation of agriculture to climate changes. This approach was critically analysed. Several neglected effects of climate change on agriculture and processes affecting agriculture's adjustment to climate change were identified. These included biological, geographical and institutional considerations. A brief account was then given of agricultural policies that governments may adopt in response to climate change. Particular attention was paid to the extent to which a laissez-faire approach is justified.

While agriculture may overcome all of the production and environmental challenges which it faces in this century, it is hard to escape the conclusion that agricultural landscapes and associated environments will change greatly in this century. The major challenge is how best to manage these changes, particularly from an economics point of view.

NOTES

1. Global population increased slowly when humans depended only on hunting and gathering and was more constrained until after agriculture commenced. The development of agriculture (and non-agricultural production which it permitted) eventually lifted the human carrying capacity of the Earth. The extent to which it can continue to do that constitutes part of contemporary debates about the sustainability of economic development.
2. It should not, however, be assumed that the activities of hunters and gatherers had no impact on the natural environment. In some early cases, they hunted species to extinction. For example, the Maoris hunted moas (large flightless birds) to extinction in New Zealand. Australian Aborigines systematically used fire to improve their hunting and gathering prospects. This changed landscapes and favoured the survival of some species while disadvantaging others.
3. It is worth noting that sometimes increased fluctuation in an environmental variable (such as rainfall) can increase the average annual yield of a crop or of livestock production (Tisdell, 1983). However, the opposite effect seems to be more common. The relationship depends on the convexity or concavity of the environmental tolerance (yield) function in the range for which these relevant environmental variables fluctuate.

REFERENCES

Adamson, D., T. Mallawaarachchi and J. Quiggin (2009), 'Declining inflows and more frequent droughts in the Murray-Darling Basin: climate change, impacts and adaptation', *Australian Journal of Agricultural and Resource Economics*, **53**(3), 343–66.
Balmford, A., R.E. Green and B. Phalan (2012), 'What conservationists need to know about farming', *Proceedings of the Royal Society B*, **279**(1739), 2714–24.
Bentham, T.G., J.A. Vickery and J.D. Wilson (2003), 'Farmland biodiversity: is habitat heterogeneity the key?', *Trends in Ecology and Evolution*, **18**(4), 182–8.
Burney, J.A., S.J. Davis and D.B. Lobell (2010), 'Greenhouse gas mitigation by

agricultural intensification', *Proceedings of the National Academy of Sciences*, **107**(26), 12052–7.

Childe, V.G. (1936 [1966]), *Man Makes Himself*, London: Collins.

Collier, P., G. Conway and T. Venables (2008), 'Climate change and Africa', *Oxford Review of Economic Policy*, **24**(2), 337–53.

Daly, H.E. (1999), *Ecological Economics and the Ecology of Economics: Essays in Criticism*, Cheltenham, UK and Northampton, MA, USA: Edward Elgar Publishing.

Ehrlich, P.R. (1989), 'Facing the habitability crisis', *BioScience*, **397**, 480–82.

Garnett, T., M.C. Appleby, A. Balmford, I.J. Bateman, T.G. Benton, P. Bloomer, B. Burlingame et al. (2013), 'Sustainable intensification in agriculture: premises and policies', *Science*, **341**(July), 33–4.

Garrido, A., M. Bielza, D. Rey, M. Inés Mínguez and M. Ruiz-Ramos (2011), 'Insurance as an adaptation to climate variability in agriculture', in A. Dinar and R. Mendelsohn (eds), *Handbook on Climate Change and Agriculture*, Cheltenham, UK and Northampton, MA, USA: Edward Elgar Publishing, pp. 420–45.

Gollin, D. (2011), 'Climate change and technological innovation in agriculture: adaptation through science', in A. Dinar and R. Mendelsohn (eds), *Handbook on Climate Change and Agriculture*, Cheltenham, UK and Northampton, MA, USA: Edward Elgar Publishing, pp. 382–401.

Haub, C. (2011), 'How many people have ever lived on Earth?', Population Reference Bureau, accessed 14 October 2015 at http://www.prb.org/Publications/Articles/2002/HowManyPeopleHaveEverLivedonEarth.aspx.

Mendelsohn, R. and A. Dinar (1999), 'Climate change, agriculture and developing countries: does adaptation matter?', *World Bank Research Observer*, **14**(2), 277–93.

Mendelsohn, R. and A. Dinar (2009), 'Land use and climate change interactions', *Annual Review of Resource Economics*, **1**, 309–32.

Phalan, B., A. Balmford, R.E. Green and J.P.W. Scharlemann (2011a), 'Minimising the harm to biodiversity of producing more food globally', *Food Policy*, **36**(1), S62–S71.

Phalan, B., M. Onial, A. Balmford and R.E. Green (2011b), 'Reconciling food production and biodiversity conservation: land sharing and land sparing compared', *Science*, **333**(6047), 1289–91.

Renfrew, C. (2013), *Prehistory: The Making of the Human Mind*, London: The Folio Society.

Ringler, C., N. Cenachhi, J. Koo, R. Robertson, M. Fisher, C. Cox, N. Perez, K. Garret and M. Rosegrant (2014), 'Sustainable agricultural intensification: the promise of innovative farming practices', *2013 Global Food Policy Report*, Washington, DC: International Food Policy Research Institute (IFPRI).

Svizzero, S. and C.A. Tisdell (2014), 'Inequality and wealth in ancient history: Malthus' theory reconsidered', *Economics & Sociology*, **7**(3), 223–40.

Tisdell, C.A. (1983), 'The biological law of tolerance, average biomass and production in a variable uncontrolled environment', *International Journal of Ecology and Environmental Sciences*, **9**(2), 99–109, reprinted in C.A. Tisdell (2003), *Economics and Ecology in Agriculture and Marine Production*, Cheltenham, UK and Northampton, MA, USA: Edward Elgar Publishing.

Tisdell, C.A. (2015), *Sustaining Biodiversity and Ecosystem Functions: Economic*

Issues, Cheltenham, UK and Northampton, MA, USA: Edward Elgar Publishing.

Tisdell, C.A. and S. Svizzero (2015), 'The Malthusian Trap and the development in pre-industrial societies: a view differing from the standard one', *Social Economics, Policy and Development*, Working Paper No. 59, Brisbane: School of Economics, The University of Queensland.

Tisdell, C.A., N. Hishamunda, R. Van Anrooy, T. Pongthanapanich and M. Arjuna Upare (2012), 'Investment, insurance and risk management for aquaculture development', in R.P. Subasinghe, J.R. Arthur, D.M. Bartley, S.S. De Silva, M. Halwart, N. Hishamunda, C.V. Mohan and P. Sorgeloos (eds), *Proceedings of Global Conference on Aquaculture 2010, Farming the Waters for People and Food*, Phuket, Thailand, 22–25 September, 2010, FAO, Rome and NACA, Bangkok, pp. 303–35.

Tscharntke, T., Y. Clough, T.C. Wanger, L. Jackson, I. Motzke, I. Perfecto, J. Vandermeer and A. Whitbread (2012), 'Global food security, biodiversity conservation and the failure of agriculture intensification', *Biological Conservation*, **151**, 53–9.

United Nations Department of Economic and Social Affairs, Population Division (2012), 'World population prospects: the 2012 revision', accessed 9 July 2014 at http://esa.un.org/unpd/wpp/Publications/Files/WPP2012_HIGHLIGHTS.pdf.

Ziska, L.H. (2011), 'Climate change, carbon dioxide and global crop production: food security and uncertainty', in A. Dinar and R. Mendelsohn (eds), *Handbook of Climate Change and Agriculture*, Cheltenham, UK and Northampton, MA, USA: Edward Elgar Publishing, pp. 9–13.

10. Marine ecosystems and global climate change: economic consequences, resilience and adjustment

10.1 INTRODUCTION

Marine areas supply humanity with many valuable ecosystem services. Costanza et al. (2014) estimated that the value of ecosystem services provided by marine areas was US$49.7 trillion in 2011, a much lower aggregate value than in 1997. They conclude that this reduced value is due to deterioration in marine ecosystems. Marine services consist of marketed, partially marketed and unmarketed ones. For example, for the most part, the provision of marine fish is marketed, marine tourism and recreational services are partly marketed, and the value of marine areas in moderating rising atmospheric temperatures, caused by elevated levels of GHGs, is unmarketed. However, rising atmospheric levels of GHGs, particularly CO_2, due to well-known anthropogenic factors, are throwing the supply of marine ecoservices into disequilibrium. As a result, the economic value and supply of these services is expected to alter drastically and spatially in coming decades. Most communities that depend on marine resources for their livelihood will have to adjust to these changing conditions. Governments will need to consider whether and when they should intervene to enable affected communities to adapt to altered marine environments, and whether to adopt policies to retard the pace of anticipated adverse changes in these environments, if such policies are available.

This chapter unfolds as follows: first, it outlines major changes in marine environments which natural scientists expect to occur as a result of elevated levels of GHG in the atmosphere and considers their general implications for the presence of living organisms, and for the supply of marine ecoservices. Secondly, attention is given to insights obtained from a Norwegian investigation of the economic impacts of GHG-induced changes in marine ecosystems. Third, GHG-induced losses of coral reefs and their potential economic impacts are discussed. Fourth, the importance for economic evaluation of taking account of the opportunity costs of the loss of coral reefs is considered as well as the need for particular attention to the dynamics of changes in marine ecosystems resulting from global warming. Proposed policies for responding to predicted changes in marine ecosystems are then examined, and this is followed by a discussion of resilience.

Why is this topic important? Globally marine ecosystems have been estimated to provide a very large percentage of the flow of the economic value of ecosystem services. Based on the estimates of Costanza et al. (2014), it can be deduced that marine ecosystems in 2011 accounted for 39 per cent (US$49.7 trillion) of the aggregate economic value of the flow of ecosystem services. However, if their estimated flow of the value of ecosystem services from tidal marshes and mangroves is added to these figures, the total value of the flow of marine ecosystems in 2011 increases to US$74.5 trillion, 59.7 per cent of the global flow. Aggregate human well-being could be severely reduced as a result of losses in these services owing to climate change. Furthermore, the potential distributional consequences of changes in marine ecosystems as a result of climate change are important because many communities (particularly in developing countries) depend heavily on marine resources for their welfare. The topic is also of significance because its discussion illustrates the challenges we face in developing adequate economic and ecological models to predict changes in the provision of marine ecosystem services and in their economic values.

10.2 ABIOTIC CHANGES IN MARINE ECOSYSTEMS DUE TO ELEVATED GHG LEVELS

Elevated levels of GHG are causing global warming of the atmosphere (mainly as a result of increased levels of CO_2) and associated physical and chemical changes in marine environments. In turn, this is altering marine ecosystems and the populations of living organisms they support. Consequently, the supply of marine ecosystem services is changing and is expected to continue to do so as atmospheric GHG levels rise. Many human communities will be worse off as a result, even though some may gain economically. Predictions about the economic and social impacts of such changes are vitally dependent on the prognosis of natural scientists about the abiotic and biotic changes to be expected as a result of climate change. In order to assess their economic impacts and to identify rational policy responses, the nature, geographical location and timing of changes in marine ecosystems attributable to GHG emissions (and other causes) need to be identified. This is a complex and formidable task. Therefore, it is not surprising that

Table 10.1   An indication of the heterogeneity of marine ecosystems

Ecosystem identifiers	Sets of ecosystems
In coastal locations	Salt marshes, mangroves, estuaries
On rocky substrates	Rocky intertidal, kelp forests, coral reefs
On soft substrates	Sandy shores, seagrass meadows, shelf sea benthos
Vast oceanic ecosystems	Pelagic, polar and ice-edge, deep sea

Source: Based on Brierley and Kingsford (2009, Table 2, p. R606).

most scientific opinions about the dynamics and spatial aspects of changes in marine ecosystems are subject to significant degrees of uncertainty. This results inevitably in economic assessments reliant on those informed opinions being subject to uncertainty, uncertainty which is compounded (as a rule) by shortcomings in economic data and analysis. While this does not mean that it is impossible to make useful economic assessments of the cost–benefit and economic impact issues raised by GHG-induced changes in marine ecosystems, their likely limitations need to be kept in mind.

The first challenging aspect to note is the considerable variety of marine ecosystems. Table 10.1 provides an initial indication of that variety. However, in practice, types of ecosystems are much more heterogeneous than indicated. For example, environmental conditions experienced by coral reefs vary with their geographical location. This results in differences in the community of living organisms which they support. Furthermore, the functioning of many of these ecosystems is interdependent and there-fore they are not independent entities.

A variety of changes in marine environments are already occurring or are expected to occur as a result of elevated GHG levels in the atmosphere. These include:

- Increasing *acidification* of marine waters induced by higher levels of atmospheric CO_2. This will have adverse consequences for the survival and growth of calciferous organisms, particularly molluscs and hard corals and to a lesser extent, crustaceans (Doney et al., 2009; Kroeker et al., 2010).
- The area of the ocean suffering serious *oxygen depletion* is predicted to increase (Stramma et al., 2012). Oxygen deficient areas become virtual dead zones and support few, if any, organisms of value to human beings.
- *Sea level rises* are expected as a result of global warming. This is because this warming expands the volume of water in the oceans and

also melts global ice cover, thereby adding to this volume. The effects on marine ecosystems depend on the speed of sea level rise; the faster and more pronounced the rise, the more difficult it is for existing ecosystems to adjust (Solomon et al., 2007, Ch. 5). Some coral reefs are likely to be drowned, and dislocation can be expected in mangroves, coastal wetlands and estuaries.

- *Changes* seem to be likely *in* ocean *circulation (currents), upwelling (vertical movements of sea water) and the vertical (heat) stratification of oceans* (Sumaila et al., 2011). Among other things, this is likely to affect the local availability of fish (see, for example, Garza-Gil et al., 2011).

- *Rising ocean temperatures* affect the distribution of marine species (see, for example, Dulvy et al., 2008). These rising temperatures can result in coral bleaching, the overgrowth of corals by algae and a likely reduction in their areal range. The abundance of fish and the presence of different species of fish spatially will also alter significantly (Cheung et al., 2010).

- *Increasing storm frequency and severity* in some regions (Solomon et al., 2007, Ch. 5) will have adverse economic consequences. For example, mariculture is likely to be adversely impacted by increased damage to fish pens, and marine-based installations and seaweed beds are more prone to destruction. Furthermore, erosion of coral reefs is likely to accelerate, thereby reducing the supply of this valuable natural resource.

All of the above-mentioned abiotic changes to existing marine ecosystems will alter their biotic components. As a result, many existing marine ecosystems (possibly most) will be replaced by new ecosystems. There will be a large-scale geographical redistribution of the maximum potential fisheries catch (Cheung et al., 2010) and alterations in the geographical distribution of marine species. However, these changes will not be instantaneous and the dynamic pathways which existing ecosystems will follow in their transition are likely to vary with diverse economic consequences. Predicting their economic consequences is difficult because ecologists are uncertain about the dynamics of the adjustment of marine ecosystems in response to climate change (Miller et al., 2010). However, present indications are that some countries will benefit economically from such changes, for instance, Norway. Others are likely to be disadvantaged (Allison et al., 2009). Modelling by Armstrong et al. (2012) indicates that Norway will make economic gains as a result of an increase in available fish stocks. On the other hand, loss of coral reefs may economically disadvantage Australia and many lower-income countries. Let us consider the economic modelling of

Armstrong et al. (2012) for Norway and then discuss the possible economic consequences of loss of coral reefs due to climate change, ocean acidification and other factors.

10.3   THE ECONOMIC IMPACT OF CLIMATE CHANGE AND OCEAN ACIDIFICATION ON NORWEGIAN FISHERIES: A CASE STUDY

Modelling by Armstrong et al.

Armstrong et al. (2012) undertook the daunting task of estimating the economic impact of ocean acidification and climate change on the aggregate gross revenues generated by the Norwegian fisheries and mariculture. They estimate this for the 100-year period 2010–2110. Because they assume that the real prices which prevailed for Norwegian fish in the period 2001–10 will continue to prevail in the period 2010–2110, their results principally depend on anticipated changes in the volume of fishing output during the focal 100-year period. Their findings rely primarily on the results of the meta-analysis conducted by Kroeker et al. (2010) of the effects of ocean acidification on marine organisms and to a lesser extent on the estimates of Hendriks et al. (2010). A problem of depending on such meta-estimates is that these estimates do not relate to particular regions and are aggregative in nature. They are, for example, not specific to the Norwegian exclusive economic zone (EEZ). Moreover, the estimates of changes in survival, growth and calcification of species relate to broad groupings of species, for example, 'all' fish, molluscs and crustaceans are considered as separate groups (see Kroeker et al., 2010, p. 1426). This ignores any variability within categories.

   Armstrong et al. (2012) assume that the composition of fish species in the Norwegian EEZ does not alter over their 100-year period. However, this is unlikely to be so. Fish species are altering their geographical location due to global warming and other effects of climate change (Cheung et al., 2010). Furthermore, the dynamics of changes in fish abundance for a 100-year period are far from certain: the relationship may not accord with the linear one assumed by Armstrong et al. in their analysis.

   In their economic analysis, Armstrong et al. limit their attention to expected changes in the supply of Norwegian (produced) seafood likely to result from ocean acidification. They estimate the changes in the streams of gross revenue for fish, molluscs and crustaceans separately and for these combined. These estimates can be regarded (in terms of economic impact analysis) as variations in the level of first-round monetary injections into the Norwegian economy. They represent a form of economic activity analysis rather than cost–benefit analysis. Estimates for best and worst case scenarios are provided for fish, molluscs and crustaceans as well as for the aggregates of their changed revenue streams. Armstrong et al. (2012) predict that the combined total change in revenue for the period 2010–2110 in the best scenario case, is an increase of 105 059 million 2010-NOK in revenue and, in the worst case, a decrease in revenue of 3926 million 2010-NOK. (At the end of June 2010, a Norwegian kroner equalled US$0.152262).[1]

Allowing for Uncertainty and the Discounting of Economic Values

It could be argued on the basis of Laplace's principle of insufficient reason that these end-values and all those values in between are equally prob-able. In this case, the probability distribution of changes in total revenue is rectangular and the expected change in the gross revenue obtained in Norwegian fisheries and aquaculture is an increase of 50 566.6 million 2010-NOK for the period 2010–2110. Furthermore, given this probability distribution, the likelihood (0.06) of a negative effect is very low. Therefore, it is highly likely (probability 0.94) that, in aggregate, Norwegian fisher-ies and aquaculture will generate increased revenue as a result of ocean acidification.

Armstrong et al. (2012) also provide discounted values for these changed revenue streams but they do not explain their rationale for choosing the discount rate. It is unclear in this context whether the discounting is assumed to indicate the time preference of the current generations or the coefficient of concern of current generations for future generations.

The Norwegian results are heavily influenced by the fact that the bulk of Norway's marine food production and revenue is obtained from fish, with only a small proportion coming from molluscs and crustaceans. Moreover, if Laplace's principle is adopted, their estimates reveal that the expected changes in revenue from mollusc and crustacean production are both negative, but small in relation to the expected increase in revenue from fish production. Nevertheless, given Laplace's principle of insufficient reason, it is highly probable (probability 0.80) that revenue generated by mollusc production will fall. The likelihood of a negative change in revenue from crustacean production is less than this, namely 0.55.

Note that even if the results of Armstrong et al. are reliable for Norway, it cannot be concluded that all countries will obtain increased revenue from marine food production as a result of ocean acidification. For example, in Australia's case, crustacean and mollusc production accounts for a high proportion of the revenue generated by its marine food production.

Furthermore, it is very likely that Australia's coral reef-based fisheries will be adversely affected. Moreover, coral reef deterioration due to global climate change may have an adverse impact on Australia's revenues generated by tourism.

The Validity of Ecological Assumptions

A problem in undertaking economic analysis of this type is that ecologists are uncertain about the impacts of climate change on the biological composition of ecosystems and the dynamics of changes in these. For example, according to the modelling of Cooley et al. (2012), the lead times before mollusc productivity is adversely affected by climate change could be very long and are predicted to vary considerably between countries. Their lead time is the length of time between 2010 and the year in which aragonite saturation is predicted to fall below a critical level. They predict that virtually no reduction in mollusc production will occur until a 'transition decade' is reached in which aragonite saturation of sea water begins to fall below a critical level. Aragonite is a form of calcium carbonate and is used by molluscs to produce their shells. Taking 2010 as the base year, their lead times vary from a minimum of 14 years for Ecuador to a maximum of 45 years in North Korea. This implies that Ecuador's transition decade would begin in 2024 whereas that of North Korea would not start until 2055. The lead times predicted for Australia and the USA are respectively 17 and 18 years and imply that their transition decade would start in 2027 and 2028 respectively.

How accurate these predictions are is difficult to tell. They indicate, however, that some countries would have to prepare for a reduction in their mollusc production at an earlier stage than others. Furthermore, the nature of the impact on the productivity of molluscs predicted by Cooley et al. (2012) differs from that assumed in the analysis of Armstrong et al. (2012) because the analysis of Armstrong et al. is based on the prediction that mollusc productivity declines at a constant rate from 2010 onwards.

An additional limitation of current ecological knowledge about the consequences for marine ecosystems of GHG emissions is that it is based on partial analysis. For example, the consequences of GHG emission for marine organisms, including their location, depend on the multiple effects (listed in the previous section), see for example, Ainsworth et al. (2011). Ocean acidification is just one of these effects. For the purposes of economic evaluation, the combined multiple effects of GHG emissions need to be taken into account, not just one. For example, in the Norwegian case, what is the likelihood that fish species in its EEZ will alter and what is the probability that the primary production of available food for fish (see

Ainsworth et al., 2011) will increase or decrease due to changing ocean currents, or the increased rate at which ice sheets in the Arctic are melting? None of these factors are allowed for in the analysis by Armstrong et al. (2012).

The discussion of the Norwegian case study of the consequences of GHG emissions for marine productivity should alert us to the following: some regions and marine-based industries are likely to make economic gains, whereas others will experience economic losses as a result of global environmental changes induced by rising levels of GHG. Furthermore, the state of ecological modelling makes it very difficult to predict confidently the timing and spatial patterns of change in marine productivity as a result of global warming. This compounds the difficulty of obtaining reliable estimates of the economic impact of global climate change. In addition, different economic and ecological modelling by Eide (2007) and by Lam et al. (2014) indicate that the economic benefits from changes in marine fisheries in the Arctic are likely to be lower than the study by Armstrong et al. (2012) suggests.

10.4   CORAL REEFS, CLIMATE CHANGE AND ECONOMIC LOSS

Costanza et al.'s View of the Changing Economic Value of Coral Reefs due to Climate Change

Costanza et al. (2014, Table 3) argue that the value of ecosystem services per ha provided in 2011 by coral reefs was by far the highest for any biome. They also supply data to support their view that this biome recorded the greatest reduction in the annual economic global value of ecosystem services provided by any biome comparing this supply in 2011 to that in 1997. This is so if their hypotheses are accepted that: (1) the global area of coral reefs more than halved in this period (and that their vacated areas were replaced by seagrass and algal beds); and (2) the estimated real price per ha of coral reefs in 2011 provides a more suitable basis for estimating the change in the total global value of this biome than does the use of 1997 real unit values. Their method involves valuing ecosystem services in both 1997 and 2011 either by the real unit values prevailing in 1997 or those prevailing in 2011. Dollar values for 1997 and 2011 are adjusted to the value of the US dollar in 2007. Note that they attribute the reduction in the global economic value of coral reefs mainly to a reduction in their area as a result of coral bleaching caused by global warming.

They find that there was a reduction in the total value of coral reef ecosystem services provided in 2011 compared to 1997 of US$10.9 trillion. This was only offset to a relatively small extent by the estimated increase in the total value of seagrass and algal biomes. Although the real value per ha of ecosystem services provided by seagrass and algal biomes was lower than for coral areas, their total real economic value rose by US$1.0 trillion per year. This was because of their increased areal extent based on the assumption that they replaced depleted coral reef areas. Consequently, the decline in the net amount of the aggregate annual economic value of ecosystem services attributed to reduced coral cover is US$9.9 trillion, given the estimates of Costanza et al. (2014).

A Different Estimate to that of Costanza et al. of the Changing Economic Value of Coral Reefs due to Climate Change

The way in which Costanza et al. (2014) estimate changes in the aggregate value of ecosystem services provided by biomes is heavily influenced by their choice of whether to determine those based on 1997 real values per ha or to do so based on 2011 real values per ha. They consider whether they should apply 1997 real values to both of their focal years or 2011 values in order to estimate these changes. They favour the latter. They state 'Given the more comprehensive unit values employed in the 2011 estimates, this is [these give] a better approximation than using the 1997 unit values, but certainly still are conservative estimates' (Costanza et al., 2014, p. 156). However, they do not address the question of whether it would be even more appropriate to use 1997 real unit values for valuing the ecosystem services provided in 1997 and the real unit values prevailing in 2011 for valuing those services supplied in 2011. If this procedure is used, different conclusions follow from those of Costanza et al. Given this alternative approach, the aggregate economic value of the ecosystems services generated by coral reefs in 2011 is much higher than in 1997, when these values are expressed in terms of 2007 US dollars. This is so despite the area of coral reefs being more than halved. This is because the increased value of coral area ecosystem services per unit of their remaining area more than compensates for the reduction in these areas, given the data used by Costanza et al. This result is plausible on economic grounds, but depends on how economic value is measured.

   Costanza et al. estimated that, based on 2007 real values, the value of ecosystem services provided per ha by coral reefs was $8364 in 1997, but in 2011 this rose to $352249. The 2011 figure is 42 times that for 1997. On the other hand, it was estimated that the global area covered by coral reefs in 2011 was 0.45 of that in 1997. The relative decrease in this area is far outweighed by the comparative increase in the estimated value of coral reefs per ha. Consequently, according to these estimates, coral reefs yielded annual ecosystem services of much greater aggregated global value in 2011 than in 1997. How could this be so? One possibility is that the data is faulty but there are also other possible economic reasons for this result. Consider some of these.

Explaining the Increased Economic Value of Coral Reefs Despite their Reduced Area due to Global Warming

Two different and often conflicting approaches to economic valuations can be found in the literature. One is based on economic activity or impact analysis and the other relies on traditional economic welfare analysis (see Chapter 4 of this book). Although legitimate doubts exist about whether economic activity analysis should be regarded as a valuation technique, it is often applied as if it is, and of course it does have economic implications. While some environmental losses can increase aggregate expenditure obtained by industries reliant on affected environments, at the same time economic welfare (based on traditional economic welfare analysis) can fall (Tisdell, 2012). Take the loss of coral reefs as an example.

Both the value per ha of marketed commodities and of non-marketed ones associated with coral reefs probably increased between 1997 and 2011. In the case of marketed commodities, two factors may have been at work. The demand curve for important commodities associated with coral reefs most likely shifted upwards, for example, particularly the demand for tourism services. Secondly, given normal demand curves, the reduced global area of reefs would have increased their scarcity value and forced up the prices of marketed commodities which are complements of reef availability. If the demand schedules for these commodities are mainly in their inelastic range, total revenue associated with the dwindling global area of coral reefs increases, as illustrated in Tisdell (2015a).

Furthermore, estimates of the non-marketed economic values per ha of remaining reefs may rise as their total global area of coral reefs declines. Again, two factors can contribute to this: the sum of marginal valuations of the supply of non-marketed goods associated with the increased area of coral reefs can be expected to decline with an increase in their global area; and the collective marginal evaluation of this supply may shift upwards, as explained in Tisdell (2015a). Particularly in developed countries, the non-marketed economic values of coral reefs are likely to exceed their marketed values (Stoeckl et al., 2014).

Costanza et al. (2014) claim that it is desirable (in order to underline the value of natural ecosystems) to estimate the expenditure which would be generated by non-marketed goods if they could be marketed, and include this in estimates of their economic value. This is a type of 'proxy' expenditure approach, as explained in Tisdell (2015b, Ch. 16).

In all the above circumstances, the expenditure approach (whether based only on actual expenditures or on these plus 'proxy' expenditures) to valuing the availability of coral reefs can result in their total estimated economic value rising as their global availability declines. Consequently, those communities fortunate enough to have some areas of surviving coral reefs can have increased incomes from their available reefs. Nevertheless, using standard economic welfare analysis, global economic welfare is likely to be reduced.

Aspects of this type of global impoverishment are illustrated in Tisdell (2015a). However, in undertaking this evaluation, it cannot be assumed that areas that lose their coral reef cover will become completely devoid of economic value. Costanza et al.'s modelling assumes that areas that become devoid of coral reef cover are replaced by seagrass and recognizes that seagrass beds have a much lower economic value per ha than coral reef cover. Because of its assumed dichotomous nature of a shift from one ecosystem to another, the ecological basis of this modelling is suspect (see section 10.6).

It is worth noting that per ha economic values of ecosystems estimated by Costanza et al. (2014) and by de Groot et al. (2012) are the averages of marginal values. They are based on meta-values calculated for selected (limited) reef areas. This raises two potential problems. Possibly, on the whole, these economic values have only been estimated for the most valuable reefs. If so, the estimates will overstate values per ha for coral reefs in aggregate. This is a potential benefit transfer problem (Rolfe and Windle, 2012). A second problem arises given that the values used in these studies are marginal values because mean marginal values are not mean average values. Multiplying marginal values per ha of an ecosystem by its total area will only provide a reliable guide to its aggregate economic value if marginal and average values of the ecosystem are a constant function of its area and equal. This is effectively assumed in the above-mentioned studies. However, the above economic analysis implies that this assumption is unlikely to be satisfied. As mentioned before, Costanza et al. (2014) suppose that coral reef areas (and other ecosystem areas) lost in the period between 1997 and 2011 transited entirely from one type of ecosystem to another. This assumption is ecologically suspect.

Biological Resilience of Coral Reefs

Scientific evidence that some coral reefs can show considerable ability to recover from coral bleaching events should not be ignored, even though the diversity of their coral assemblages tends to be reduced following such events (Adjeroud et al., 2009; Roff et al., 2014; Mumby et al., 2011). We cannot rely on comparative static analysis to assess satisfactorily the economic consequences of changes in marine ecosystems. The trajectories (dynamic paths) of ecosystem change must be accounted for, taking into account, for example, the type of issues raised by Hoegh-Guldberg et al. (2007). Several of these issues are discussed in Tisdell (2015a). The analysis of Costanza et al. (2014) does not pay enough attention to the dynamics of ecosystem change because it only makes comparisons between two years. Furthermore, they suppose (at odds with the ecological findings mentioned above) that coral reef areas damaged by global warming transited entirely to seagrass ecosystems. Consequently, economic resilience issues are not considered.

10.5   OPPORTUNITY COSTS, CORAL ECOSYSTEM LOSS AND CHANGES IN ECONOMIC VALUE AND WELFARE

Most studies of the economic consequences of the loss of coral reefs are limited to the reduction in the economic value of these reefs alone. They often do not take account of the value of ecosystems that replace them. Usually these replacements are not worthless. Secondly, and more importantly, they fail to take account of the economic value of anthropogenic activities which are the drivers of such losses. In other words, they ignore opportunity costs. When all these economic values are taken into account (despite a fall in the economic value of coral reefs due to their loss), the aggregate economic value of economic and environmental changes associated with this loss can be, but need not be, positive. This can be so even though the allocation of resource use or utilization of ecosystems is not ideal from a collective economics point of view. This is not to suggest that no serious economic deficiencies exist in human use of natural ecosystems. They do because of the existence of market failures and other identifiable problems. However, it is irrational and unscientific when assessing the economic consequences of a change in a particular ecosystem to focus only on the alteration in *its* economic values. Consider this matter in more detail.

As pointed out above, Costanza et al. (2014) provide evidence to indicate that the global economic value of coral reefs fell by US$10.9 trillion between 1997 and 2011. On the other hand, they assumed that seagrass and algal biomes replaced the areas vacated by coral reefs and that the global value of this biome increased by US$1.0 trillion. Consequently, the net change in the value of the area once occupied by coral reefs was

US\$9.9 trillion. Apart from whether or not their assumption about the dichotomous nature of marine ecosystem change is ecologically accurate, this figure of US\$9.9 trillion does not measure the change in the aggregate value of human activities and developments associated with this change.

Loss of coral reefs can be attributed to multiple factors of which an elevated level of GHG in the atmosphere is just one. Terrestrial developments, such as agriculture extension and intensification and growing urbanization in watersheds that drain into coral reef systems, also take their toll on reefs because nutrient levels and sediment increase in reef areas as a result of water run-off. These are considered to be major obstacles to conserving the Great Barrier Reef, even though they are not the only anthropogenic problems of concern. However, it is undeniable that terrestrial economic developments have economic value. Consequently, the main issue which needs to be addressed is whether the extra economic value obtained as a result of terrestrial development and activity exceeds any economic loss which it generates by altering marine ecosystems.

Figure 10.1 provides a crude representation of the relevant economic

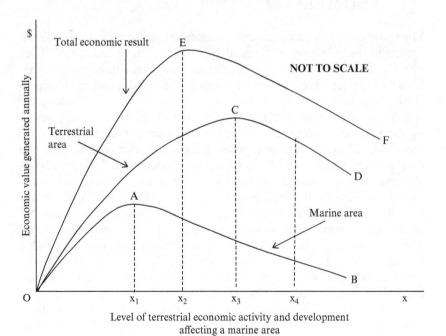

Figure 10.1	*An illustration of the importance of taking into account*
			opportunity costs when evaluating the economics of changes
			in marine ecosystems

problem. There the relationship OAB represents the net annual economic value generated by a marine area in relation to the level of terrestrial economic activities and developments. The marine area may, but need not, consist entirely of the same biome. The terrestrial activities could include those increasing the emission of GHG. In this particular case, it is assumed that up to level x_1, there is complementarity between the economic value of the marine area and terrestrial economic development and activity. The relationship OCD represents the annual economic value generated as a result of terrestrial activity and development which impacts on the environment of the focal marine area. OEF indicates the total annual economic value obtained from the focal marine area plus that flowing from terrestrial economic activity and development. Note that the drawing is not to scale but is constructed so as to highlight the critical relationship.

It is assumed that neither the relevant marine area nor the terrestrial area contributes economic value in the absence of terrestrial economic development. With increasing terrestrial economic activity and development, the flow of the economic value from the terrestrial area and the associated marine area both increase until x equals x_1. After that, due to negative environmental spillovers from terrestrial development, the economic value of benefits from the marine area begins to decline. However, the economic value obtained from terrestrial activity and development continues to increase until x reaches the level of x_3. The combined flow of economic value obtained from this relevant terrestrial area and the marine area reaches a maximum for the level of terrestrial activity and development of x_2. At this point, the marginal increase in the value generated by terrestrial activity and development equals the marginal reduction in the economic value extracted from the focal marine area.

If the relationships in Figure 10.1 represent social economic values, then the optimal degree of terrestrial economic activity and development is x_2. Compared to the situation prevailing at x_1, it is socially optimal to degrade the focal marine area to some extent. However, market and political failures may result in terrestrial activity and development being substantially in excess of x_2 (for example, at x_4), resulting in economic welfare being lower than is attainable. The extent to which this is happening is the major economic issue which needs to be addressed in considering environmental change. It is uneconomic to adopt a policy which maximizes the economic value of the flow of ecosystem services from a marine area by treating it as an independent entity. This is because to do so ignores opportunity costs.

The illustration in Figure 10.1 assumes a steady state situation. Consequently, it ignores sustainability issues and the dynamics of changes in ecosystems. The dynamics of ecosystem changes (particularly of marine ecosystems) in relation to stresses are complex and uncertain. This creates serious complications for the economic valuation of the consequences of climate change. Let us consider briefly some aspects of the dynamics of marine ecosystem changes and their consequence for economic valuation.

10.6   THE DYNAMICS OF CHANGES IN MARINE ECOSYSTEMS AND THEIR CONSEQUENCES FOR ECONOMIC VALUATION

In order to satisfactorily determine the economic consequences of changes in ecosystems, particular attention needs to be paid to the trajectories of these changes. These trajectories are usually more complex and uncertain than is assumed in contemporary economic assessments of these changes (for example, that of Costanza et al., 2014). For instance, coral reefs seem to be more resilient to heat stress than is commonly supposed.

The Adaptability of Marine Ecosystems to Climate Change: The Example of Coral Reef Ecosystems

Ecologists have found that not all areas of coral are equally susceptible to global warming and heat stress and that some recover from bleaching events rather rapidly. Consequently, in these areas, coral reefs continue to be the dominant ecosystems. Ecological reasons for this resistance can include the following:

1.   Coral species (or varieties) present in the area that are more resistant to heat stress multiply to replace less heat-resistant ones.
2.   Heat-resistant coral species and symbionts may migrate to the area from elsewhere.
3.   Darwinian selection may favour species and varieties which are more tolerant of heat stress.
4.   Some corals may exhibit genetic plasticity in response to heat stress.
5.   To some extent this epigenetic factor is Lamarckian in nature. It implies that some corals undergo genetic transformation in response to heat stress and that the genetic inheritance of their offspring makes them more resistant to heat stress than their parents.

The above factors can influence the trajectories of the presence of coral reefs and their economic consequences. Moreover, the time at which each of these factors is likely to become a major influence on the composition of species in a reef area can be expected to vary. For example, factors 1 and 2 may dominate prior to factors 3 and 4. It is therefore probable that the biological diversity of some coral reef areas will initially decline in response to heat stress (that is, following bleaching) and then subsequently increase in biodiversity compared to their nadir. Consequently, their economic value may change in a similar pattern. This is not to deny that some areas of coral reefs may be replaced 'permanently' by different ecosystems.

Scientific evidence exists to support the hypothesis that coral reef systems can exhibit considerable adaptability to climate change, although the composition of coral reef species may alter. For example, Adjeroud et al. (2009) found that coral assemblages in Moorea, French Polynesia showed significant resilience following one cyclone and four bleaching events between 1991 and 2006. Although the presence of turf algae increased considerably following these events, it was unable to dominate the reef system for very long. The reef returned to its pre-disturbance coral cover within a decade. However, the composition of the coral community changed. There was an increase in the presence of *Porites*, the cover of *Acropora* returned to pre-disturbance levels but the presence of *Montipora* and *Procillipora* was reduced. Therefore, this coral reef exhibited a high degree of resilience but changed its composition of coral species. Similarly, Roff et al. (2014) found that populations of *Porites* corals in the Rangiroa Atoll, French Polynesia, showed remarkable ability to recover from mass coral bleaching despite initial populations experiencing significant partial mortality. The offspring of more heat-resistant numbers of *Porites* which survived the bleaching event rapidly re-colonized the skeletal remains of members succumbing to bleaching.

In the above cases, it seems likely that reef resilience is primarily due to Darwinian selection in situ. Furthermore, in some cases, persistence may be achieved by the migration of coral species that show greater resistance to heat stress or their symbionts (Ortiz et al., 2014). Moreover, epigenetic changes may occur in some corals that increase the adaptability of their populations to climate change. Mumby et al. (Online Appendix, (2011) state, in this regard: 'It is known or assumed that sublethal stress induces temporal gene expression that allow cells to acquire cytoprotection or more permanent structural changes to enable organisms to survive stressful conditions that are normally lethal.'

Of course, the above scientific findings do not imply that coral reef systems are impervious to climate change. They do, however, imply that the responses can be quite varied. Undoubtedly, some coral-dominated systems can be replaced permanently by an algal-dominated system due to environmental changes attributable to climate change and due to the adverse effects of regional anthropogenic stresses, such as elevated levels of nutrients and sediments in sea water caused by land-based economic activities. Therefore, an existing marine ecosystem may change from one type to another. Consider some of the dynamics of such adjustments and their implications for economic valuation.

The 'Flipping' of Marine Ecosystems to New Equilibria

Climate change and other stressors are liable to throw coral-dominated ecosystems into disequilibrium. One can imagine two alternative possible stable equilibria for areas currently dominated by coral reefs. If stress is kept below some threshold they may remain in that state, but should it reach high enough levels they may gravitate towards algal-dominated systems (Hoegh-Guldberg et al., 2007, p. 1739). Even if global environmental change cannot be controlled, it may be possible by managing other stressors to prevent or delay the transition of a coral-dominated area to an algal-dominated one. For example, Hoegh-Guldberg et al. (2007) suggest that allowing the population of reef-grazing fish to increase in parts of the Caribbean (by reducing the catch of fish that consume algae), can delay or prevent coral reefs from becoming algal-dominated for some time. However, they do not explore the economics of adopting such a policy.

Depending upon where coral reefs are located, other stresses can be potentially reduced to lower the speed at which areas of coral reefs stressed by climate change can alter to algal-dominated ones. For example, the run-off of water from terrestrial economic activities (such as agriculture) in the Great Barrier Reef region stresses coral in some areas and favours algal growth. This is because it increases the level of nutrients and sediments in several areas where corals now exist. In other words, these terrestrial activities generate negative environmental externalities because of their adverse impacts on coral reefs. The economics of policies to reduce these stresses are considered later in this chapter.

It was noted above that, in some circumstances, hard corals show considerable ability to recover from events commonly associated with global warming. This ought to be given greater attention in considering economic policies to cope with global warming. The trajectories of changes in marine ecosystems generated by climate change are important for determining several aspects of economics and economic policy. Different trajectories have very different economic consequences, change economic valuations and alter the nature and timing of appropriate economic policies. For example, the modelling of Armstrong et al. (2012) of the effect of ocean acidification on mollusc productivity differs from that of Cooley et al. (2012). Armstrong et al. suppose that the process has already begun whereas Cooley et al. are of the view that it will not occur until several years hence, with highly variable lead times between countries. Consequently, it may be many years before some countries experience reduced mollusc production as a result of ocean acidification. Given this time delay, invest-ment and re-investment in mollusc production is likely to remain profitable for several years yet in many countries. The economic length of life of many mollusc-aquaculture projects if commenced now will be less than the number of years before increased acidification impacts on mollusc produc-tion. Therefore, it would be irrational not to proceed with these projects in the short term if they are profitable.

The Sensitivity of the Value of Norway's Fisheries to Assumptions about Changes in Marine Ecosystems

The flow of economic value from Norway's fisheries estimated by Armstrong et al. (2012) is very sensitive to the predicted pattern of its fish productivity. For example, Armstrong et al. predict a linear relationship like ABC in Figure 10.2. However, it might be that the actual relationship turns out to follow a path like ABDE. In that case, factors associated with climate change initially increase the level of Norway's fish production but subsequently they cause it to decline. Changes in Norway's fish production will not only be influenced by alterations in ocean acidification, but prob-ably also by factors such as changes in currents, in ocean upwelling, ocean

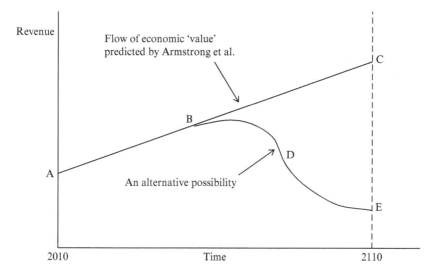

Figure 10.2   When all the effects of climate change (not only acidification) are taken into account, the impact on the aggregate receipts obtained by Norwegian fisheries may follow a different pattern from that predicted by Armstrong et al. (2012)

warming and variations in the rate of melting of Arctic ice cover. This highlights the limitation of only considering the economic impact of just one environmental change associated with global warming. The combined effects of these changes can also be expected to change the composition of wild fish species in many cases, and different species may have different economic values (Sumaila et al., 2011).

Many different trajectories for the adjustment of marine ecosystems to climate change can occur and these will have different implications for the economic values associated with these changes. Armstrong et al.'s estimates for Norwegian fisheries are based on a continual increase in the economic value of Norway's fisheries' production. The following is a sample of other possible changes in the economic value of a marine area as a function of time in response to climate change:

1. This value may at first increase and then subsequently decline to reach a higher level than initially.
2. This value may initially increase and then decline to a level lower than initially.
3. It may initially decline and then increase to a level higher than originally.
4. It may fall at first and then increase to a level less than in the first place.
5. This value may initially increase then plateau at a higher level than in the beginning.

Economic Valuation of Different Trajectories of the Flow of Economic Value

It is difficult to know how best to value the economic worth of the flow of ecosystem services. However, in case 1 above, it is clear that changes in the flow of economic value as a result of ecosystem change compared to the initial state are superior to persistence of the initial state. In case 5, while the flow of value as a result of ecosystem change always remains higher than if the initial state persists, it does reach a maximum and then shows some decline. Therefore, according to one criterion of sustainability (which requires economic value to never decline), it is considered to be undesirable (Tisdell, 2015b, Ch. 4; and Chapter 3 of this book). However, sustainability criteria would reject all the other cases as undesirable compared to this initial state.

If the capitalized values of alternative paths are used to determine the desirability of alternatives, then both cases 1 and 5 would be rated as more desirable than the initial state. Case 4 would be rejected as less desirable. However, depending on the pattern of the trajectory, and the discount rates applied to future flows, cases 2 and 3 may be rated as more desirable than the initial state.

10.7   ECONOMIC POLICIES FOR RESPONDING TO CLIMATE-INDUCED CHANGES IN MARINE ECOSYSTEMS

Now consider economic policies which may be adopted for responding to climate change. Two types of policy can be considered:

- policies to retard ecosystem change (mitigation strategies);
- policies that are based on the inevitability of ecosystem change and that are designed to cope with it (adjustment strategies).

In some cases, these policies may be pursued in conjunction. For example, ecosystem change may be retarded but it may still be necessary to adopt policies for economic and social adjustment to the changing ecosystem. Let us consider economic aspects of the first mentioned type of policy response initially and then pay attention to the second type of policy response.

Economic Aspects of Policies to Slow Changes in Marine Ecosystems

It is generally accepted that even if there are stresses on marine ecosystems caused by global climate changes, it may be possible to retard changes in these systems by reducing local anthropogenic stresses. The possibility of doing this depends on locational factors and the type of marine ecosystem being placed under stress. The deterioration of coral reefs in some locations may, for example, be retarded by reducing the run-off of nutrients and sediment from nearby land areas where run-off is a function of anthropogenic activity. This is a policy option being pursued for helping to conserve parts of the Great Barrier Reef. If, for example, fertilizer use in agriculture is identified as an important contributor (via nutrient run-off) to coral reef stress, then from an economics point of view, consideration should be given to reducing its use in the relevant area until the marginal economic benefit of improved reef quality equals the marginal loss in economic benefit from agriculture.[2]

In countries such as Australia, where coral reefs are spread geographically over very large areas, it is unlikely to be economic to adopt spatially uniform measures to conserve reef areas affected by human activity. Appropriate measures will depend on relative economic values. For example, if in the *most* northerly region of the Great Barrier Reef, the economic value of intensified agriculture is low relative to the loss in economic value it causes to coral reefs, this weakens the case for encouraging the intensification of agriculture in this northern area.

The above analysis may also be applied to changes in other marine ecosystems. For example, increases in nutrient loads and sediments in the China Sea contribute to the increased frequency and the extent of red tides in this area and these have adverse impacts on the fisheries located there. Therefore, China needs to consider in what areas and to what extent priority should be given to reducing nutrient and sediment discharges into the China Sea.

Spatial issues of the above kind have received little attention from economists. However, given that it is unlikely to be economic to conserve existing marine ecosystems everywhere, they are important. There is no escaping the fact that it is desirable to consider economic trade-offs in making decisions about the conservation of ecosystems.

Adjustment Policies Based on the Assumption that Changes in Marine Ecosystems are Inevitable

Some changes in marine ecosystems seem to be inevitable. That raises the question of what policies, if any, should be adopted by governments to respond to the elements of climate change that they do not individually control. This is a complex matter because the timing of ecosystem changes due to climate change and their nature are still subject to considerable uncertainty.

Views about the extent to which governments should intervene in adjustments to climate change can differ substantially. On the one hand, the view may be adopted that, at the very most, only limited government intervention is called for. Particularly if environmental change is slow, communities may adjust adequately to it of their own accord. At the other end of the spectrum is the belief that governments should play a leading role in preparing individuals and communities for such changes and mandate that particular precautionary actions be taken.

The timing of adjustment to climate change can be problematic. In terms of timing, some adjustments to climate change may be premature and in other cases not made early enough. For example, (as mentioned above) there may be projects for which the length of their economic life is such that if they are proceeded with now, they will give a sufficient return to make them worthwhile before the full impact of climate change adversely affects them. However, at a later time, these projects may no longer be economic given the increased immediacy of significant adverse environmental impacts from climate change. For instance, some types of mariculture may no longer be economic once impacts from climate change become significant. Prior to this they may still be economic even if some impacts from climate change are apparent.

Those who favour minimal government intervention in the economy might argue that the government ought to limit itself to providing public goods to facilitate adjustment to climate change and other commodities that cannot be adequately supplied by markets. The supply of public goods could include the supply of information about the expected nature and timing of climate change. However, there is the problem that information provided by governments is not always reliable. Furthermore, there may be a case for governments undertaking works to provide local public goods that delay some consequences of climate change, for example, in some cases building or raising dykes to respond to sea level rises.

To some extent, private insurance markets can be expected to respond to changes in the risks associated with climate change. For example, insurance premiums for mariculture operations facing increasing risks as a result of climate change can be expected to rise. This should send a market signal to mariculturalists that adjustments are called for.

10.8   ECONOMIC AND ECOSYSTEM RESILIENCE AND ALTERATIONS TO MARINE ECOSYSTEMS CAUSED BY GLOBAL CLIMATE CHANGE

Economic resilience in response to climate change has become a subject of increasing interest. However, the term 'resilience' is frequently used without paying much attention to its meaning. It usually refers to the ability of an entity to recover or bounce back to a previous state (or approach it) following dislocation by an adverse event. Basically, it is a dynamic concept.

Two aspects of resilience are relevant to the present discussion, namely:

- the ability of marine ecosystems to recover from the adverse impacts of climate change; and
- the ability of socio-economic systems to do likewise.

A limitation of the concept in this context is that it tends to indicate implicitly that the biological and economic consequences of climate change are always adverse. However, as the cases considered in this chapter demonstrate, this is unlikely to be so in all localities.

The concept of resilience has achieved prominence because it has come to be regarded as the most important feature of sustainability of ecological and economic systems. In reality, however, it is only one attribute of such systems which needs to be considered when assessing their performance if they are subjected to environmental changes, such as climate change (Tisdell, 2015a). A few relevant aspects of resilience can be mentioned here.

First, economic resilience can be increased by adopting precautionary and anticipatory actions. However, this can be done too early from an economics point of view. For example, as mentioned previously, although ocean acidification seems likely to eventually have adverse consequences on the cultivation of molluscs (such as oysters), a significant amount of time may expire before this effect occurs or becomes an economic concern (Cooley et al., 2012). In many cases, no changes in current practices will be economic for some time to come.

Secondly, the need for economic resilience in relation to changes in marine ecosystems depends to a large extent on resilience of these systems or, more generally, the dynamics of their responses to climate change. Some marine ecosystems in some situations may exhibit a limited amount of resilience. There is, for example, evidence of partial recovery of coral reefs from heat stress (Roff et al., 2014). However, the effect of stresses on marine ecosystems is complex. It is likely to depend on several dynamic features, for instance, whether the stresses are sudden, gradual, intermittent, and rising in magnitude. Also, the combination of stressors is important.

Despite all these difficulties of scientific prediction, current evidence indicates that there is a high probability that countries in low latitudes (nearest the equator) will suffer the most adverse economic consequences from changes in marine ecosystems due to climate change whereas those in higher latitudes are likely to gain (Sumaila et al., 2011; Cheung et al., 2010; Allison et al., 2009). Consequently, tropical, less-developed countries (which depend heavily on living marine resources) are likely to be most adversely affected economically by these changes. This is especially so for most island countries. Furthermore, the capacity of these countries to cope with these changes is extremely limited (Tisdell, 2008). In fact, compared to most higher-income countries, lower-income countries have limited scope and capacity to adjust to major threats to the livelihoods of their citizens which are expected as a result of anticipated alterations to their marine environments arising from climate change. In other words, their scope and capacity for economic resilience in these circumstances is severely constrained. In many cases, substantial environmental deterioration is likely to trigger increased migration from less-developed countries, and many island countries may no longer be inhabitable (Tisdell, 2008).

10.9   CONCLUDING COMMENT

The discussion in this chapter supports the view of Stoeckl et al. (2011, p. 126) that in order to devise appropriate economic policies for the management of ecosystems, 'we do not just need information about total and marginal values [of the services they provide] but we also need information on the social, temporal and spatial distribution of these values'. This is especially important in the case of ecosystems that change their nature in response to alterations in environmental conditions, such as those associated with elevated levels of atmospheric GHG. It is also desirable that these valuations be based on sound and explicit economic theory and pay adequate attention to the state of ecological knowledge, including its limitations.[3]

NOTES

1. The Norwegian kroner estimates have not been translated into US dollars because the rate of exchange is quite variable and alters considerably over long periods. Furthermore, the argument is unaffected by this.
2. Note that this is a very simplified policy perspective. It ignores the transaction costs involved in policy implementation, and fertilizer use is not the only contributor to nutrient run-off generated by the practice of agriculture.
3. Preparation of this chapter has benefited from helpful information supplied by Peter Mumby and Sabah Abdullah of the School of Biological Sciences, The University of Queensland, John Rolfe of the University of Central Queensland, Natalie Stoeckl of James Cook University, Rashid Sumaila of the University of British Columbia, and Jyothis Sathyapalan of the Centre for Economic and Social Studies, Hyderabad as well as suggestions by Karachepone Ninan. This help is appreciated.

REFERENCES

Adjeroud, M., F. Michonneau, P.J. Edmunds, Y. Chancerelle, T.L. de Loma, L. Penin, L. Thibaut, J. Vidal-Dupiol, B. Salvat and R. Galzin (2009), 'Recurrent disturbances, recovery trajectories, and resilience of coral assemblages on a South Central Pacific reef', *Coral Reefs*, **28**(3), 775–80. DOI: 10.1007/s00338-009-0515-7.

Ainsworth, C.H., J.F. Samhouri, D.S. Busch, W.W.L. Cheung, J. Dunne and T.A. Okey (2011), 'Potential impacts of climate change on Northeast Pacific marine foodwebs and fisheries', *ICES Journal of Marine Science: Journal du Conseil*, **68**(6), 1217–29. DOI: 10.1093/icesjms/fsr043.

Allison, E.H., A.L. Perry, M.-C. Badjeck, W.N. Adger, K. Brown, D. Conway, A.S. Halls, G.M. Pilling, J.D. Reynolds, N.L. Andrew and N.K. Dulvy (2009), 'Vulnerability of national economies to impacts of climate change on fisheries', *Fish and Fisheries*, **10**, 173–96.

Armstrong, C.W., S. Holen, S. Navrud and I. Seifert (2012), 'The economics of ocean acidification – a scoping study', *Fram Centre*, accessed 30 June 2015 at http://www.surfaceoa.org.uk/wp-content/uploads/2012/05/The-Economics-of-Ocean-Acidification.pdf.

Brierley, A.S. and M.J. Kingsford (2009), 'Impacts of climate change on marine organisms and ecosystems', *Current Biology*, **19**, R602–R614.

Cheung, W.W.L., V.W.Y. Lam, J.L. Sarmiento, K. Kearney, R. Watson, D. Zeller and D. Pauly (2010), 'Large-scale redistribution of maximum fisheries catch potential in the global ocean under climate change', *Global Change Biology*, **15**, 24–35.

Cooley, S.R., N. Lucey, H. Kite-Powell and S.C. Doney (2012), 'Nutrition and income from molluscs today imply vulnerability to ocean acidification tomorrow', *Fish and Fisheries*, **13**, 182–215.

Costanza, R., R. de Groot, P. Sutton, S. van der Ploeg, S.J. Anderson, I. Kubiszewski, S. Farber and R.K. Turner (2014), 'Changes in the global value of ecosystem services', *Global Environmental Change*, **26**, 152–8.

de Groot, R., L. Brander, S. van der Ploeg, R. Costanza, F. Bernard, L. Braat, M. Christie, N. Crossman, A. Ghermandi, L. Hein, S. Hussain, P. Kumar, A. McVittie, R. Portela, L.C. Rodriguez, P. ten Brink and P. van Beukering (2012), 'Global estimates of the value of ecosystems and their services in monetary units', *Ecosystem Services,* **1**(1), 50-61. DOI: 10.1016/j.ecoser.2012.07.005.

Doney, S.C., V.J. Fabry, R.A. Feely and J.A. Kleypas (2009), 'Ocean acidification: the other CO_2 problem', *Annual Review of Marine Science*, **1**(1), 169–92. DOI: 10.1146/annurev.marine.010908.163834.

Dulvy, N.K., S.I. Rogers, S. Jennings, V. Stelzenmüller, S.R. Dye and H.R. Skjoldal (2008), 'Climate change and deepening of the North Sea fish assemblage: a biotic indicator of warming seas', *Journal of Applied Ecology*, **45**, 1029–39.

Eide, A. (2007), 'Economic impacts of global warming: the case of the Barents Sea Fisheries', *Natural Resource Modeling*, **20**(2), 199–221. DOI: 10.1111/j.1939-7445.2007.tb00206.x.

Garza-Gil, M.D., J. Torralba-Cano and M. Varela-Lafuente (2011), 'Evaluating the economic effects of climate change on the European sardine fishery', *Re gional Environmental Change*, **11**(1), 87–95. DOI: 10.1007/s10113-010-0121-9.

Hendriks, I.E., C.M. Duarte and M. Alvarez (2010), 'Vulnerability of marine biodiversity to ocean acidification: a meta-analysis', *Estuarine, Coastal and Shelf Science*, **86**, 157–64.

Hoegh-Guldberg, O., P.J. Mumby, A.J. Hooten, R.S. Steneck, P. Greenfield, E. Gomez, C.D. Harvell, P.F. Sale, A.J. Edwards, K. Caldeira, N. Knowlton, C.M. Eakin, P. Iglesias-Prieto, N. Muthiga, R.H. Bradbury, A. Dubi and M.E. Hartziolis (2007), 'Coral reefs under rapid climate change', *Science*, **318**, 1737–42.

Kroeker, K.J., R.L. Kordas, R.N. Crim and G.G. Singh (2010), 'Meta-analysis reveals negative yet variable effects of ocean acidification on marine organisms', *Ecology Letters,* **13**, 1419–34.

Lam, V.W.Y., W.W.L. Cheung and U.R. Sumaila (2014), 'Marine capture fisheries in the Arctic: winners or losers under climate change and ocean acidification?', *Fish and Fisheries*, online first. DOI: 10.1111/faf.12106.

Miller, K., A. Charles, M. Berenge, K. Brander, V.F. Galluci, M.A. Gasalla, G. Munro, R. Mutugudde, R.E. Onmer and R.I. Perry (2010), 'Climate change, uncertainty and resilient fisheries: institutional responses through integrative science', *Progress in Oceanography*, **87**, 338–46.

Mumby, P.J., I.A. Elliott, C.M. Eakin, W. Skirving, C.B. Paris, H.J. Edwards, S. Enríquez, R. Iglesias-Prieto, L.M. Cherubin and J.R. Stevens (2011), 'Reserve design for uncertain responses of coral reefs to climate change', *Ecology Letters*, **14**(2), 132–40. DOI: 10.1111/j.1461-0248.2010.01562.x.

Ortiz, J., M. González-Rivero and P. Mumby (2014), 'An ecosystem-level perspective on the host and symbiont traits needed to mitigate climate change impacts on Caribbean coral reefs', *Ecosystems*, **17**(1), 1–13. DOI: 10.1007/s10021-013-9702-z.

Roff, G., S. Bejarano, Y.-M. Bozec, M. Nugues, R. Steneck and P. Mumby (2014), '*Porites* and the Phoenix effect: unprecedented recovery after a mass coral bleaching event at Rangiroa Atoll, French Polynesia', *Marine Biology*, **161**(6), 1385–93. DOI: 10.1007/s00227-014-2426-6.

Rolfe, J. and J. Windle (2012), 'Testing benefit transfer of reef protection values between local case studies: the Great Barrier Reef in Australia', *Ecological Economics*, **81**, 60–69. DOI: 10.1016/j.ecolecon.2012.05.006.

Solomon, S., D. Qin, M. Manning, Z. Chen, M. Marquis, K.B. Averyt, M. Tignor and H.L. Miller (eds) (2007), *Contribution of Working Group I to the Fourth Assessment Report of the Intergovernmental Panel on Climate Change,* Cambridge and New York: Cambridge University Press, accessed 28 April 2017 at http://www.ipcc.ch/publications_and_data/ar4/wg1/en/contents.html.

Stoeckl, N., M. Farr, D. Jarvis, S. Larson, M. Esparon, H. Sakata, T. Chaiechi, H. Lui, J. Brodie, S. Lewis, P. Mustika, V. Adams, A. Chacon, M. Bos, B. Pressey, I. Kubiszewski and R. Costanza (2014), 'The Great Barrier Reef World Heritage Area: its "value" to residents and tourists, Project 10-2 Socioeconomic systems and reef resilience', *Cairns: Reef and Rainforest Research Centre Limited.* Final Report to the National Environmental Research Program, accessed 23 July 2015 at http://www.nerptropical.edu.au/publication/project-102-final-report-great-barrier-reef-world-heritage-area-its-%E2%80%98value%E2%80%99-residents.

Stoeckl, N., C.C. Hicks, M. Mills, K. Fabricius, M. Esparon, F. Kroon, K. Kaur and R. Costanza (2011), 'The economic value of ecosystem services in the Great Barrier Reef: our state of knowledge', *Annals of the New York Academy of Sciences*, **1219**(1), 113–33. DOI: 10.1111/j.1749-6632.2010.05892.x.

Stramma, L., E.D. Prince, S. Schmidtko, J. Luo, J.P. Hoolihan, M. Visbeck, D.W.R. Wallace, P. Brandt and A. Kortzinger (2012), 'Expansion of oxygen minimum zones may reduce available habitat for tropical pelagic fishes', *Nature Climate Change*, **2**(1), 33–7. DOI: 10.1038/nclimate1304.

Sumaila, U.R., W.W.L. Cheung, V.W.Y. Lam, D. Pauly and S. Herrick (2011), 'Climate impacts on biophysics and economics of world fisheries', *Nature Climate Change*, **1**, 449–56.

Tisdell, C.A. (2008), 'Global warming and the future of Pacific Island countries', *International Journal of Social Economics*, **35**(12), 889–903.

Tisdell, C.A. (2012), 'Economic benefits, conservation and wildlife tourism', *Acta Turistica*, **24**, 127–48.

Tisdell, C.A. (2015a), 'Marine ecosystems and climate change: economic issues', *Economics Ecology and the Environment,* Working Paper No. 199, Brisbane: School of Economics, The University of Queensland.

Tisdell, C.A. (2015b), *Sustaining Biodiversity and Ecosystem Functions: Economic Issues*, Cheltenham, UK and Northampton, MA, USA: Edward Elgar Publishing.

Index

Aborigines, Australian 14, 45, 48, 65, 82, 155
Adjeroud, M. 217
Age of Discovery 15
AgForce 123
agrarian societies 86–91
Agricultural Revolution 2, 10, 16, 85, 88, 155–6
 and social and economic change 86–8
agriculture 4, 11–12, 17, 44, 78, 86, 88, 138, 155–7, 166–7, 169, 197–9, 218, 221–2 *see also* Agricultural Revolution
 agricultural adjustment 192–6
 coping with increased demand for produce 184–7
 effects of climate change on 188–97
 and environmental change 14–15
 historical context 180–84
 related to environmental change and economic welfare 173–80
Aguilar, F.X. 110
Ainsworth, C.H. 208
air conditioning 21–3
air quality 22–3, 61, 73, 157, 163
Akerlof, G.A. 96, 115
albatross 144, 146
alcohol 82, 95, 113–14, 116
altruism 36, 54, 123–5
Anderson, R.C. 110, 113–14
animal cruelty 54
animal welfare 95–6, 107–8, 111–12, 115–16, 126
anticipated competitive behaviour 84–5
anti-social behaviour 79–80
aquaculture 12, 181, 207, 219
aragonite 208
Armstrong, C.W. 205–7, 218–20
Armstrong, K. 86–7
Arrow, K.J. 69

barriers to entry 134–5
Bentham, T.G. 186
Bergson, A. 36, 68–70
biofuels 159, 167–8, 187
biological carbon cycles 156–7
biological conservation 3, 120–21, 145
 behaviour of public organizations and NGOs in relation to 129–35
 conflicts between conservationists 144–5
 ecotourism enterprises 143–4
 koala case study 139–42
 and market systems 127–8
 motives and perceptions 123–7
 policies and valuation techniques 135–8
 role of wildlife rehabilitation centres 142–3
 topics and issues 121–3
biological diversity 121–2, 150, 217
biological tolerance functions 194–5
biomass 3–4, 155, 159, 161, 167–8, 194
birth rates 26
Body Shop 111
bounded rationality 51, 112
Bronze Age 183
brumbies 121, 124–5, 144–6

Cai, Z. 110
cap and trade systems 174
capital 73
capitalism 84
carbon footprint 163
carbon sinks 157, 167 *see also* CO_2 (sequestration of)
carrying capacity problem 158
caste system in India 90
Chen, J. 110
Childe, V.G. 12, 181
choice modelling methods 57
Ciriacy-Wantrup, S.V. 55, 125–6

Clark, C. 17
climate change 1, 3, 5, 7, 80, 149, 170
    see also global warming
    and agriculture 188–97
    biophysical processes 154–7
    combating 157–8
    consequences of for agricultural
        production 4
    and coral reefs 209–13
    economic strategies to reduce
        emissions 163–6
    and electricity generation 159–63
    failure to control emissions 150–54
    and fisheries 206–9
    and marine ecosystems 4, 216–18
    migration as a response to 26–7
    and renewable energy *see* renewable
        energy
club goods 22
CO$_2$ 3–5, 122, 167, 174, 189
    carbon sinks 157, 167
    failure to control emissions 150–54
    following Industrial Revolution
        154–6
    and marine ecosystems 203–4
    reducing emissions through
        renewable energy 159–63
    sequestration of 149, 151, 158,
        166–8
collective action 37
collective choice 67–9
committees 83–4
communities 40–41, 43–5
computable general equilibrium models
    190
conditions for sustainable economic
    development 37–9
consumer protection 97–100
consumer sovereignty 3, 94–6,
    114–17
    costs of satisfying food safety
        standards 106–7
    economic aspects of food safety
        standards 100–104
    and environmental, social and
        animal welfare 107–12
    externalities arising from final acts
        of consumption 112–14
    information asymmetry and
        consumer protection 96–100

opportunity costs and supply
        responses 104–6
consumption, extraordinary increase
    in 1
contingent valuation methods 56, 63
Cooley, S.R. 208, 218
coral bleaching 205, 212–13, 217
coral reefs 4, 122, 176, 202, 208–16,
    221, 224
    biological resilience of 212–13
Costanza, R. 202–3, 209–13
cost–benefit analysis 53–4, 177,
    179–80, 204, 207
    social cost–benefit analysis 51, 56,
        58, 62, 138
cultural embedding 80–82

Daly, H.E. 174
dams 58
De Groot, R. 212
death rates 26
decision-making 5–6, 52–3, 62, 84, 130,
    137, 151
deontological ethics 53–4, 108
desertification 192
digital technology 13
Dinar, A. 188–90, 192–3, 195
division of labour 12, 181
drought 168, 185, 189, 195

economic growth treadmill 83
economic impact analysis 51, 74–5
economic liberalism 114, 116, 151
economic multiplier effect 143
economic policies 75, 83, 116, 121,
    141–2, 149, 154, 164, 180, 187,
    196–7
    and biological conservation 135–8
    and marine ecosystems 221–3
Economic Revolutions 12–13 *see also*
    Industrial Revolution
economic valuation 2, 51–2, 56, 75,
    121, 146, 211 *see also* values
    and biological conservation 135–8
    local and regional economic impact
        analysis 74–5
    macroeconomic indicators of
        changes in human well-being
        52, 70–74
    and marine ecosystems 216–21

of unmarketed environmental
  commodities 56–67
welfare economics approaches 67–9
economic welfare 35–6, 47, 70–71, 89,
  91, 105, 173, 180, 211, 213–16
ecotourism 121, 143–4
education 71–2, 81
eggs, free-range 95, 98, 111–12, 116
Ehrlich, P.R. 2, 10, 182–3
  Ehrlich's equation 10, 18–21, 29
Eide, A. 209
elephants 126, 143, 146
Elliott, J. 110
employment 74, 83 *see also*
  underemployment; unemployment
endangered species 122, 134, 140 *see
  also individual species*
environmental Kuznets curve 10, 17,
  20–21, 29, 32, 48
environmental poverty trap 42–4
environmental spaces 18, 21–3
ethics 53–5, 62, 108, 110–11 *see also*
  values
exotic species 120, 125–6, 178

family sizes 16
farming *see* agriculture
feral animals 121, 124–5, 144–5
fertilizer 176, 178, 184, 188, 221
First Economic Revolution 12, 181
fisheries 4, 44, 205–9, 219–20, 222
flood control 58
flooding 43–4, 188
flow resources 168–9
Food and Drug Administration 100
food safety standards 94, 98–107, 115
food waste 187
foresight intelligence 150–54
Forest Stewardship Council 109–10
fossil fuels 3, 12, 14, 21, 149, 152, 155,
  157–8, 160, 163, 166, 170
  economic strategies to alter use of
    163–6
freedom of choice 94
future generations 33–5, 38–9, 47, 207
  and CO_2 emissions 152–3
  concern of current generations for
    the welfare of 36–7
  sacrifice of current generations of
    behalf of 35–6

Galbraith, J.K. 114
GDP 17, 70–71, 113
genetic engineering 196
genetic resources 120–21
geological carbon cycles 156–7
geothermal power 159–61
global warming 1, 3–4, 6, 21–2,
  149–51, 154–6, 170, 173 *see also*
  climate change
  and the Industrial Revolution 16
  and marine ecosystems 203, 211–12,
    220
  measuring the impacts of 188–92
  naturally caused 24–6
globalization 103, 113, 128, 183
GNI 70–72
governance 41, 84, 136
governments 98–100, 116, 121, 128,
  131, 139, 150, 164, 197, 202–3, 222
  see also economic policies
Gowdy, J. 78, 89, 91
Great Barrier Reef 122–3, 214, 218,
  221–2
greenhouse gases (GHGs) 1, 3, 5,
  80, 149, 153–4, 157 *see also*
  CO_2
  economic strategies to reduce use of
    163–6
  failure to control emissions
    150–54
  and the Industrial Revolution 16
  and marine ecosystems 202–6,
    208–9, 214–15, 225
Greenpeace 131

Hamilton, S.F. 101, 111
Hammoudi, A. 104
Hartwick, J. 38
Haub, C. 183
Hayek, F. 94–6, 100, 114
heat stress 189–90, 194, 216–17
hedonic pricing methods 61–2
Hendriks, I.E. 206
historical perspective on environmental
  change 2, 9–11, 29–30
  Ehrlich's equation and
    environmental consequences
    18–21
  increased exploitation of natural
    resources 27–8

socio-economic consequences
 of major changes in natural
 environments 24–7
stages of economic development 2,
 10–18
value of man-made environments
 21–4
Hobbes, T. 153
Hoegh-Guldberg, O. 213, 218
horses, wild 121, 124–5, 144–6
Human Development Index (HDI)
 70–72
human well-being 52, 70–75, 107–8,
 115–16, 174, 203
hunting and gathering 10–12, 82, 89,
 91, 155, 181–2, 198
 and environmental change 14
Husson, S.J. 138
hydrocarbons 158, 162, 164–5, 167,
 169–70
hydropower 159–62

income inequality 10, 23–4, 29
income levels 73–4, 88
Industrial Revolution 2, 10–12, 24, 90,
 157, 181, 183, 198
 CO_2 accumulation following 154–6
 and environmental change 15–17
 and social and economic change
 88–9
industrialization 15–16, 23, 25, 88, 113
 see also Industrial Revolution
information and communication
 revolution 13
information asymmetry 94, 96–100,
 108
information overload 112
information technology 89
interest groups 129–31, 145
International Monetary Fund (IMF)
 130
irrigation 15, 44, 178, 184–5, 188, 193

Jackson, T. 165
job security 89

Kaldor–Hicks criterion 67–8, 126, 136,
 144–5, 177, 179, 193
Kant, Immanuel 53, 55, 108, 125
koalas 134, 139–42, 146

Koenig, A. 101–2
Krall, L. 89
Kroeker, K.J. 206
Kuznets, S. 23
Kyoto Protocol 151

laissez-faire approach 198
Lam, V.W.Y. 209
Laplace's principle of insufficient
 reason 207
large numbers problem 153–4
leakage 74
legislation 122–4 *see also* economic
 policies
Leopold, Aldo 55, 126
let the buyer beware 97
life expectancy 35, 45, 71–2
livestock, grazing 175, 177–8, 186–8
living standards 157

macroeconomic indicators of changes
 in human well-being 52, 70–74
Malthus, T.R. 11, 87
man-made dams 58
man-made environments 10, 21–4
manufacturing 12, 17, 181, 183
marine ecosystems 4, 176, 202–3, 225
 abiotic changes in 203–6
 coral reefs 4, 122, 176, 202, 208–16,
 221, 224
 dynamics of changes in 216–21
 economic and ecosystem resilience
 223–4
 fisheries 4, 44, 205–9, 219–20, 222
 market failure 3, 95–100, 115, 127–8,
 145, 175, 197
 market reforms 38, 82
 market systems 79, 81–3, 90, 94–5,
 127–8, 130, 145, 151, 164, 174,
 197
 and biological conservation 127–8
 economic policies for 221–3
materialism 13
McCluskey, J.J. 101
McDonald's 111
Mendelsohn, R. 188–90, 192–3, 195
metallurgy 182–3
migration 10, 13, 16–17, 24, 45–6, 89,
 224
 of minority groups 45

as a response to environmental change 26–7
Milankovitch cycles of global cooling and warming 155
Millennium Ecosystem Assessment 120
minimum viable population concept 55
minority groups 45, 48, 65, 82, 155
moral hazard 166
moral responsibility 84–5
Mumby, P.J. 217

neoclassical economics 51, 57, 95, 117
neo-Malthusianism 89
noise pollution 61, 73
noisy-signal problem 153–4
non-government organizations 3, 110, 121, 127, 130, 139–41, 145
functions performed by 131–2
possible 'efficiency failures' of 132–5
normative economic models 52
nuclear power 160, 168–9

ocean acidification 3–4, 150, 156, 169–70, 204, 206, 208–9, 219–20
Olson, M. 80
one-child policy 20
on-farm environmental changes 175–6
opportunity costs 104–6, 137–8, 203, 213–16
orangutans 55, 125, 138, 143, 146
Oxfam 111
oxygen depletion 204

Paretian improvement criterion 67–8, 126, 136, 144–5, 177, 179, 193
Pareto, V. 95, 127
Paris Agreement 151, 170
Paris Convention on Climate Change 32
Parker, C. 111–12
Pascual, U. 57
pastoralism 11, 27
Pearce, D. 36–7, 152
pest control 122, 125–6
pesticides 101, 104–6, 178, 184
photosynthesis 156–7
Pigou, A. 52, 56
Pigovian tax 164
Polanyi, K. 78, 89
political lobbying 163, 167

pollution 32, 61–2, 73, 157, 163, 169
population levels 1, 4, 9, 11–13, 20, 87–8, 156, 158, 165–6, 168, 177, 181–4, 186, 188
and the Industrial Revolution 16–17
population mobility 45–6
population trap 87
positivist economic models 52
Posner, Richard 127
poverty 48, 88, 165
environmental poverty trap 42–4
and sustainability issues 33, 42–5
pressure groups 130, 137
prisoner's dilemma problem 79–80, 149, 152
privatization of natural resources 46
product chains 95, 98, 103–4, 115–16
property rights 81, 167, 174, 177
public goods 10, 22, 145, 223

quality of life 11, 72
quality standards 115 *see also* consumer protection
quasi-public goods 22

rainfall 25, 168, 192–3, 197
rainforests 15
Rawls, John 33–4, 47, 127
Rehm, J. 113
religion 13, 81–2, 90
renewable energy 159–63
biofuels 159, 167–8, 187
biomass 3–4, 159, 161, 167–8, 194
geothermal 159–61
hydropower 159–62
nuclear 160
power 168–9
solar 3, 158–60, 162–3, 168
tidal power 159
wind 3, 14, 158–63, 168
revealed preference methods 57–62, 101
Ricardian method 190, 198
Ringler, C. 185–6
Roff, G. 217
Rolfe, J. 111
Ross, L. 80, 150–54
Roy, A.D. 34
rule of law 41

Schelling, T.C. 152
Schlömer, S. 159–62, 169
sea level rise 24, 150, 154–6, 170,
    188–9, 204–5
Second Economic Revolution *see*
    Industrial Revolution
sediments 122, 222
self-interest 51, 79–80, 83, 95, 122–3,
    127–8, 130, 138, 149, 177
service sector 17, 88
Smith, Adam 11–12, 95, 127
social capital 41
social cohesion 41, 48
social cost–benefit analysis 51, 56, 58,
    62, 138
social embedding 1–3, 5–7, 78–9, 91–2,
    128, 135–6, 145
    cultural embedding 80–82
    and failure to control CO$_2$ emissions
        150–54
    impact of changing economic
        systems on 85–91
    structural embedding 79–80
    types of in modern economies
        82–5
social structures 85
social values 81–2
soil erosion 175
solar energy 3, 158–60, 162–3, 168
species extinction 14, 17, 141, 145
stages of economic development 2,
    10–18
standard of living 16, 181
Starbucks 111
starvation 15
stated preference methods 57, 62–7,
    101, 136
steady state economy 165
Stoeckl, N. 225
structural lock-in 152
structural social embedding 79–80
subsidies 162–3, 166
superorganisms 85
sustainable (economic) development 2,
    46–9, 74
    definition of 32–3
    different criteria for the occurrence
        of 33–7
    different indicators of 40–42
    and population mobility 45–6

poverty and 42–5
proposed conditions for achieving
    37–9
three pillars required for 39–40, 42,
    47–8
Svizzero, S. 78, 87
systematic bias in stated values 66

taxes 162–4, 166
technological path dependence
    152
tertiary sector 17, 88
Third Economic Revolution 13
tidal power 159
Tisdell, C.A. 32, 34, 55, 75, 78, 80, 85,
    87, 138–9, 141, 143, 152, 186, 211,
    213
tourism 22, 58, 61, 121, 142, 202, 208
    see also travel cost methods
    ecotourism 121, 143–4
trade unions 83
Traill, W.B. 101–2
transaction costs 80, 97, 128, 196
transparency 53
travel cost methods 57–61
trilogy concept 39–40, 42, 47–8
Trump, Donald 165
trust 41, 98, 115
Tscharntke, T. 186
tyranny of large numbers 84

ultrasociality 6, 85, 89–91
uncertainty effect 136
underemployment 72
unemployment 45, 72, 83, 89
UNESCO 122
United Nations 130–31, 184
United Nations Conference on
    Environment and Development
    (UNCED) 32
United Nations Development
    Programme (UNDP) 70–71
unmarketed environmental
    commodities 56–67
urbanization 11–12, 16, 182
utilitarianism 125

values 2, 51–2, 81–2, 207–8, 211–16,
    220–22, 225 *see also* economic
    valuation

of conserving natural biodiversity versus consequential approaches 55
different types of 52–4
of unmarketed environmental commodities 56–67

water power 14
weather forecasting 197
Webster, R. 165
Weisdorf, J.L. 78
welfare economics 52, 67–9
welfare functions 69–70
wildlife rehabilitation centres 121, 142–3

willingness-to-pay 54, 61–7, 101–2, 105, 108, 110, 115–16
Wilson, C. 143
wind energy 3, 14, 158–63, 168
wishful thinking 153–4
World Bank 130–31
World Commission on Environment and Development (WCED) 33
World Trade Organization (WTO) 130–31
World Wildlife Fund (WWF) 123, 131–2

zero population growth 165
Ziska, L.H. 189